THE IMPERIAL STUDIES SERIES

THE STAPLE TRADES OF THE EMPIRE

BY VARIOUS WRITERS

EDITED BY

ARTHUR PERCIVAL NEWTON

Elibron Classics
www.elibron.com

Elibron Classics series.

ISBN 1-4021-4170-X (paperback)
ISBN 1-4021-0146-5 (hardcover)

This Elibron Classics Replica Edition is an unabridged facsimile of the edition published in 1918 by J. M. Dent & Sons, Ltd., London and Toronto.

STAPLE TRADES OF THE EMPIRE

THE IMPERIAL STUDIES SERIES

THE STAPLE TRADES OF THE EMPIRE

BY VARIOUS WRITERS

EDITED BY

ARTHUR PERCIVAL NEWTON

M.A., D.Lit., B.Sc.

LECTURER ON COLONIAL HISTORY IN THE UNIVERSITY OF LONDON,
UNIVERSITY AND KING'S COLLEGES

MCMXVIII

LONDON AND TORONTO

J. M. DENT & SONS, LTD.

PARIS; J. M. DENT & FILS

A series of Imperial Studies Lectures delivered in the London School of Economics and Political Science in 1917.

CONTENTS

THE STAPLE TRADES OF THE EMPIRE

INTRODUCTION

The Lectures that are here collected were delivered within the University of London at the London School of Economics and Political Science in the Spring of 1917. The course was planned under the auspices of the Imperial Studies Committee of the University and the Royal Colonial Institute, but for the views expressed in his lecture, each lecturer alone is responsible, and the task of the Committee has been solely to ensure that each trade should be dealt with by some one who could speak with authority concerning it. In a certain sense the present volume is complementary to that preceding it in the Imperial Studies Series on *The Exploitation of Plants*, edited by Prof. F. W. Oliver, F.R.S. In that volume the actual production of certain commodities is dealt with, while in these lectures attention is mainly directed not to the raising of the products, but to their distribution from the producer to the consumer through the processes of commerce. The merchants are necessarily closely concerned with production and with an extension of the area from which they draw their merchandise, so that questions such as these must necessarily form a part of the matter dealt with, but looked at from a different point of view than by those whose business is concerned with actual production.

Among the many tasks that have fallen to the universities in the course of the war, one that has been of not less importance than the others has been the provision of lectures to the general public on matters of vital interest to the Empire, both in the political and the material sphere. The English universities have done great work for the advancement of knowledge, but it has been one of the severest reflections upon them that they have been somewhat removed from the problems of the work-a-day world and have had less direct influence than has been exercised by the universities of some other advanced countries. The responsibility for this has been avowedly due in part to the people in our academic world who have had too much a tendency to keep themselves cloistered within their studies and laboratories, but it is even more to be attributed to the English public at large, who have regarded their universities merely as glorified scholastic institutions where young men might finish their schooling amid pleasant surroundings before they went forth to take up their real work in the world. The foundation of newer universities in the great centres of population, which began in the first half of the nineteenth century with the setting up of University College and King's College in London and Owens College in Manchester, marked the beginning of a fresh and less cloistered type of institution, and the process has gone on at an accelerated pace in the course of the last thirty years. The new colleges have come into close contact with the problems of the great urban communities, and men are beginning to realise better that their universities are not merely places for professional education, but have an even more important function to fulfil in the life of a progressive nation as factories and clearing-houses of ideas. The working out and the translation of ideas into effective form, if undertaken by properly trained minds and according to exact scientific methods, supplies a large part of what is called "Research," and it is best undertaken not by individuals working in isolation, but by bands of professors, assistants and students working in harmony in a university centre. There is little doubt that, urged on with a new zest for knowledge in the period of reconstruction after the war and in the new age that will follow it, these methods of collective team-work in research will be used to a greatly increased degree, and that in their universities men will apply the patient processes of scientific investigation to many subjects which as yet have hardly been touched. The new work to be dealt with will be found in almost every field of knowledge and not least in those fields that especially concern the progressive political and material life of the Empire in which we live.

Among these Imperial Studies a place of considerable importance must be allotted to those that relate to economic development, and particularly to a study of the conditions that govern the trades in the great staples of commerce whereby the Empire provides its subsistence and a considerable proportion of its material wealth. Such a study undertaken without proper contact with other branches of knowledge would inevitably lead to false conclusions and to too great an insistence upon purely material agencies in human progress. It cannot be denied that the history of the past, looked at from the point of view of economic and commercial factors, is of immediate utility for those who wish to mount to the sources of the great streams of action of the human race and to search for the connection one with another of the happenings of our own time. Much first-rate work has already been done in the study of the past on these lines, but comparatively little has been done in an examination of the great activities of commerce at the present day, and a vast amount of spade-work has to be done before a comprehensive view of the sources of our imperial wealth can be obtained. The purely economic historian, like the doctrinaire economist, has a tendency to take the view that throughout the ages material interests have held a preponderant place in the march of civilisation; and during the period that immediately preceded the outbreak of the world war, this view was held to an exaggerated degree, and it appeared as though material considerations of self-interest weighed with decisive effect upon the destinies of states. But the war has opened men's eyes to the partial nature of the view. We now realise that though material interests set things running and are exceedingly potent owing to the continuous nature of their action, yet when they tend to become predominant, they bring into play other and more deeply seated forces that are usually forgotten or minimised because they are quiescent. At great crises the more powerful spiritual motives of nationality are set free. They deflect the line of action of material interests with tremendous effect; men and nations cease to be swayed solely by the desire for gain, and the resultant action is in a profoundly different direction from what would have been anticipated if the economic forces alone had been taken into account.

The surface appearance of things changes, but the great abiding facts of human nature remain. The generations succeed one another with qualities and defects almost identical, and despite the different conditions of their environment men obey the same social and national tendencies as those who preceded them. To-day, as centuries ago, the great organised peoples employ their activity in the exploitation of regions that are abandoned or undeveloped, obeying those hereditary tendencies that urge on the human race to take possession in its entirety of the domain that nature has assigned to it.

One difference alone separates earlier times from our own, the revolution that has been brought about by the great discoveries in the realm of natural science in the course of the past two centuries. This has enabled us to appreciate more fully than our fathers the riches and the resources of the globe, and has provided us with more perfect tools for their utilisation and application. But however perfected these tools may be, they have only increased power when the thinkers who direct their application have benefited by a study of the experiences of the past, and by careful study have realised the present conditions in which they have to be used. Unless the past is studied, the lessons of experience cannot be acquired; unless present circumstances are carefully surveyed it is impossible to apply the lessons of the past with profit. In our personal lives we know that the same faults engender the same crises, and the same principles, whether good or bad, lead to the same consequences; but in the national and imperial sphere the appreciation of this fundamental axiom is perhaps hardly so clear. If in spite of the repeated lessons of history the men of to-day still err, it is because that presumptuousness which is always closely allied to ignorance is an essential human frailty and because, perhaps, more than any other of the great peoples, Englishmen have preferred to guide their actions by their ingrained prejudices, to stick to the counsels of the "practical man" and to distrust the advice of the trained thinker.

The lectures that are here printed may afford an illustration of the fact that the best "practical man" may also be a scientific thinker about the conditions of the trade to which he has devoted his energies, and may supply within the very narrow limits assigned a survey of the actual conditions of certain great branches of commerce at the present day. The trades dealt with are those in Wheat, Wool, Cotton, Oils and Fats, Metals and Sugar, and in each case the trade has been dealt with by an expert who can speak with first-hand knowledge. Each lecturer has dealt with his subject not merely from the historical standpoint, but also with the design of indicating the conditions that will govern future development and of emphasising the steps that in his opinion are necessary to secure the greater benefit of the Empire as a whole. In a democratic community where broad questions of principle are decided by the voice of public opinion, if the decisions are to be sound, it is essential that the general mass of citizens shall be provided with reliable data concerning the issues to be dealt with. In making such data accessible the universities can play an exceedingly important part by the provision of courses of public lectures delivered by experts of acknowledged weight in an atmosphere from which party propaganda is banished. When, as in the present case, such lectures have been delivered, it is the further task of those who have arranged them to make them accessible to a wider public by their publication in collected form. A book such as this, that can safely be commended to the general reader, in no sense completes the work of the university in the field, but should lead on to the publication of further materials of all kinds ranging from the highly technical and specialised monograph embodying the results of original research to the great synthetic treatise weaving those researches into a connected whole. For the publication of works such as these, only the endowments of a properly organised University Press can provide, and in the University of London, the richest city of the world, such endowments are entirely lacking. It is, therefore, a matter of congratulation when, owing to the enterprise of a publishing house, it is possible to secure the publication even of the popular survey.

Though the treatment of the commercial questions here dealt with has been entirely independent in each case, it will be noted what a considerable unanimity of opinion is evinced by the lecturers as to the direction in which future efforts must be made if the British Empire is to recover in full strength from the losses of the war and of the period of slackness and easy-going indifference that preceded it.

In the words of the Royal Commission on the Resources of the Dominions—

> "It has been a commonplace for years that British manufacturers and merchants should be stimulated to study and cultivate the Dominion markets, and to produce goods of the nature and quality which are required by consumers across the seas. It has equally been a commonplace that the manufacturers of the Mother Country should be urged to use the vast resources of raw materials which the Dominions possess. The difficulty has lain in suggesting acceptable measures for the realisation of these ideals. Hitherto the proposals made have been mainly the work of advocates of some particular fiscal or other theory, which they have pressed, in season and out of season, as a universal remedy. In our judgment these counsels, however important they may be, cover only a part of the problem. An Imperial policy, in the broadest sense, must include much that is not fiscal. There is no short cut to the formulation of such a policy; what is needed is detailed examination of existing conditions, and practical and definite proposals for the removal of difficulties and for securing co-operation."

The present volume is a small contribution to that task of detailed examination.

ARTHUR PERCIVAL NEWTON.

University of London.

OILS AND FATS IN THE BRITISH EMPIRE

By SIR A. D. STEEL-MAITLAND, Bart., M.P.
His Majesty's Under-Secretary of State for the Colonies

To deliver an address on oils and fats from which petroleum and other mineral oil products were excluded would, before the war, have appeared very surprising. The importance of vegetable and animal oils and fats was then in general only partially understood. Now, however, the problem of the blockade of Germany, as well as that of the provision of adequate supplies for this country, has caused it to be fully realised. Indeed the list of the principal uses of oils or oil products only needs mention for their indispensable character to become patent. In human diet oils are becoming more and more an essential part. The growth in the consumption of margarine has been enormous, and it absorbs a large proportion of the nut oils imported. For other edible purposes, particularly in frying and cooking, cotton-seed oil is employed in large quantities, and the odour is familiar to nearly every Londoner. Lard and olive oil, for the latter of which ground-nut oil is a very general substitute, are equally well known. Among non-edible uses the fatty acids derived from oils are the main constituents of both soap and candles. On the importation of linseed depends the supply of paint in the building and decorating trades, as also the manufacture of linoleum. The provision of milk and of home-grown meat is largely determined by the supply of an adequate quantity of agricultural feeding cake, which is the by-product formed when the oil is expressed from the oil seed or oil nut. Lastly, glycerine, obtained formerly as a by-product from oil in the manufacture of soap, is of quite peculiar importance at the present moment. While high explosives are principally formed from mineral sources, glycerine is requisite for British propellant explosives, and the enormous quantities required in the war has made the provision of adequate supplies at once a problem of absolute necessity and of considerable difficulty.

The Principal Oil Nuts, etc.

Before dealing in detail with the commercial uses of the various oils, and the economic position of this and other countries in regard to them, it may be well to enumerate them briefly and to describe the processes of manufacture which are common to them. Vegetable oils are obtained from nuts, seeds, or beans. Coconut oil, which is perhaps the most important of all, is derived from copra, which is itself the white substance found within the ordinary coconut. The chief sources of supply are India, Ceylon, the East Indies and the Southern Pacific. The oil palm, which is a West African tree, produces a nut which yields two oils, which are very different in character: palm oil from the fleshy exterior or "pericarp," and palm-kernel oil from the kernel of the nut. Ground-nut oil, produced from the ground nut, pig nut, or aracheide nut, as it is variously called, comes principally from West Africa and India. It is a soft oil, that is to say, liquid at ordinary temperatures, in contradistinction to the hard oils previously mentioned, of which the melting point is from 80° Fahrenheit and upwards. Of seeds the most important are linseed, cotton seed, rape seed and sesame seed. Among beans the soya bean and the castor oil bean are of chief importance for oil production.

THE PROCESS OF CRUSHING, ETC.

When the nuts, seeds, or beans, as the case may be, have been brought to the factory, they are either passed through a press and the oil crushed out, or—a less common process, at any rate, in this country—a solvent, such as benzine or trichlorethylene, is added after a preliminary crushing and the oil is subsequently recovered. In either case two products emerge: the oil and the feeding cake or meal. The process of crushing produces cake, and that of extraction, as the latter is called, yields a lighter and less compressed substance in the form of meal. Both this and the cake are to all intents and purposes in their final commercial form. It is only necessary to call attention here to two points. First, that very varying proportions of oil and residue respectively are obtained from different seeds or nuts. Thus it is calculated that the average yield will vary from as much as 60 per cent. of oil from copra to not more than 12 per cent. or 13 per cent. in the case of soya beans.[1] The proportions of cake, of course, vary inversely. The above are *average* yields, but the percentage obtained varies to some extent with the machinery used for pressing. In this country, where, prior to the war, there was very little crushing of nuts as distinct from seeds, the common type of machinery was the Anglo-American cage press which served very well in the case of soft seeds like cotton or linseed. It will readily be understood, however, that such machinery was not fitted to get the best results from nuts or kernels which are much harder substances, and at the present time only a few firms have installed the up-to-date machinery which is necessary to deal with these satisfactorily. The difference in the yield of oil may be considerable, and amounts in the case of

[1] The yield or ratio for converting from seed into oil for all seeds is given in Col. 1 of the following table—

—	Yield of Oil	Digestible Food Units in Cake		Digestible Food Value in cake from 1 Ton of Seed
		Board of Agriculture Table	Linseed = 100	Linseed = 100
	1	2	3	4
Linseed	33⅓	120	100	100
Cotton-seed *	16⅔	68·6	57·17	71·43
Rapeseed	33⅓	—	—	—
Soya beans	13	122·3	102	127
Copra	60	102·6	85·5	51·3
Palm kernels	45	90·5	75·4	62·2
Ground-nuts	40	145·2	121	108·9
Sundry seeds	33⅓	—	—	—
Maize †	—	94	78·7	—

* Undecorticated cotton-seed has been taken. A mean has also been taken between the figures for Egyptian and Indian seed.

† Maize is only added for a comparison of its nutritive qualities.

palm kernels to as much as 5 per cent. In other words two and a half more tons of oil will be obtained from 100 tons of kernels by the new machinery than by the old. Even, therefore, in times when palm-kernel oil does not reach the present price of £55 per ton, the economy is considerable. By the extraction process the yield of oil is yet greater, only about 1 per cent. to 1½ per cent. of oil being left behind. On the other hand, oil so produced presents greater difficulties to the refiner, while the resultant feeding stuff is neither so nutritive nor so easy to handle.

The second point to be borne in mind is that apart from this difference between cake and meal, feeding cakes produced from the different seeds differ among themselves in nutritive value, and in the presence or absence of other qualities not susceptible of quantitative measurement. Not only so, but their respective market values are also largely affected by the degree to which the farming community is familiar with them. The table given on p. 12 shows the digestible food units of the different cakes, and from this an approximate idea can be obtained of their food values.

The Processes of Splitting, Refining and Hydrogenation

The manipulation and uses of oils are more complicated than those of feeding cakes, and any survey should include animal fats, such as tallow or whale oil, in the case of which, of course, no preliminary process of crushing or extraction is necessary, nor is there any by-product in the shape of feeding stuff. Two things may happen to the oil. Either it may be employed in its whole form, as, for example, by the margarine maker or linoleum manufacturer, or the oil may be separated or "split" into its constituent parts of glycerine and fatty acids. In nature this separation is caused by the action of an enzyme or ferment in the oil, and the rancidity with which most persons are familiar, is nothing else than the state of the oil after the fat-splitting enzyme has been at work. In modern manufacture an analogous result is obtained by artificial methods. Prior to the war the glycerine was simply a by-product from the fatty acids, which are the staple article in the manufacture of soap and candles. The position is now reversed. The demands of the Ministry of Munitions for glycerine are enormous. It is, in fact, now the main product, and in view of the fact that oils, as a rule, contain not more than 10 per cent. of glycerine, the disposal of the fatty acids becomes a very difficult problem, especially in view of the fact that they cannot easily be stored for an indefinite period. Experiments have been made within the writer's knowledge in the direction of enabling fatty acids to be used in place of the unsplit oil in the manufacture of linoleum, but unfortunately the attempt to find in them a satisfactory "boiled oil substitute," by which attractive title that category of substances is known, has hitherto been unsuccessful. The process of splitting and the production of glycerine have been mentioned in view of their importance at the present time. The largest outlet, however, is for the whole oils, the next stage in the preparation of which is, therefore, refining. But on this, though it is perhaps the most important process in the manufacture, a very few words must suffice, for it is here that the most jealously guarded secret chamber of the trade lies, into which no outsider may penetrate. What is certain is, that the chemical side of the treatment of oils, as expressed more particularly in their refining, is of the utmost importance, and it is with regard to it that the greatest possibilities of development lie. It is by additional refining that oils may be raised, so to speak, in the scale of uses, and rendered edible, where before they were only available for other purposes. Parallel with improvements in refining is the discovery and application of the process of "hydrogenation," or increasing their content of hydrogen. By this means the melting-point is raised, and they are thus again rendered applicable to new uses. This has been done in the case of whale oil, which is now utilised as a "hard" oil for soap manufacture, while certain Continental factories are said to have also refined it up to the point of making it suitable even for margarine. The importance is thus emphasised of giving due scope and recognition to the chemist's work, as was undoubtedly done in Germany. While, however, it is fair to say, as the result of a tolerably close study and comparison of evidence as to methods there and in the United Kingdom respectively, that certain firms here have appreciated and acted on this fact, it remains true that, taken as a whole, manufacturers in this country have failed to let the chemical side of the management permeate and affect the working of the concern to the degree that is really necessary if constant progress is to be maintained.

THE SPECIAL USES OF THE DIFFERENT OILS

The question which naturally arises out of the brief list of the main outlets for oil given at the beginning of this paper, is that of the special purposes to which each oil is suitable. The curious and complicating position—and in this, indeed, lies a large part of the interest of the whole subject—is that it is practically impossible to give a simple and straightforward answer to such a question, the fact being that for every use enumerated, with but few exceptions, one oil is interchangeable with others under certain conditions of price. On the other hand, it is probably true to say that the only two oils which are mutually replaceable for all the purposes to which they are put, are coconut oil and palm-kernel oil. The result of this is, that oils generally form, as it were, a chequers puzzle, and manufacturers will substitute one for another, according to the supplies of the different kinds available at given prices, any manufacturer of one product, however, being affected in his choice of pieces by the prices paid by other classes of manufacturers. Even so, fairly specific groups can be distinguished into which various oils may be placed according to their different qualities, although fresh discoveries as to ways of treating them chemically may, as already indicated, at any time result in their addition to another group.

Thus coconut oil and palm-kernel oils are hard oils, *i. e.* oils which melt at a high temperature. They are principally used for margarine, which consists of hard oils to the extent of 57½ per cent. to 67½ per cent., according to the time of year, the proportion being greatest in summer. The fatty acids from these oils are also used for the higher classes of toilet soap in view of their lathering qualities.

Between the hard and the soft oils proper come tallow and its substitutes, of which fish oil is the principal. These are used in nearly all soaps, though in different proportions. Another intermediate oil, which may be said to stand between the tallows (with which it is to some extent interchangeable) and the soft oils, is palm oil.

The soft oils include a greater variety. For edible purposes cotton-seed oil and ground-nut oil are the principal, and form 15 per cent. to 25 per cent. of margarine (according to the season), apart from their uses for frying or for salad.

The soap manufacturer has a catholic taste, and can utilise practically every kind of oil. For soft soap, which is in a class by itself, he uses linseed oil in its natural state, and also whale oil and cotton-seed oil. For hard soaps linseed and whale oil both require hydrogenating, and are then combined with the other oils to form the cleanser soaps (Sunlight soap, Crosfield's Perfection, etc.), or with palm-kernel oil or coconut oil (usually the former) for toilet soaps.

The demand of the linoleum trade is confined to linseed oil, for which in this connection no substitute has been found, and the same is true of the paint and varnish trades (except that the former also uses a certain quantity of soya oil). For lubricating, the soft oils, such as that from rape seed or unhydrogenated fish oil, are used, while rape oil is also used for burning.

The above description of the principal uses of the different oils is given in a condensed form in the following table. In addition to the purposes mentioned there, however, the splitting of oils for glycerine creates a wide demand for certain oils, those which are principally utilised for this purpose being linseed oil, whale oil, and, to a lesser degree, palm oil.

	Burning	Lubricating	Edible	Paint	Varnish	Linoleum	Soap
Linseed oil . .	×	..	..	×	×	×	×
Cotton-seed oil .	..	..	×	..	..	..	×
Soya oil . . .	×	..	×	×	..	..	×
Rape oil . . .	×	×	..	..	..	..	..
Coconut oil . .	..	..	×	..	..	..	×
Palm-kernel oil .	..	..	×	..	..	..	×
Ground-nut oil .	..	..	×	..	..	..	×
Palm oil . . .	..	..	..	..	..	..	×
Fish oil . . .	×	×	..	..	..	..	×
Tallow . . .	..	..	×	..	..	..	×

In determining a choice between different oil-bearing nuts or seeds it will be remembered that, in addition to the oils themselves, the comparative values of feeding stuffs also form an important consideration. This fact, both from the national point of view, when supplies are restricted by tonnage difficulties or from the manufacturers' point of view in calculating his total return, may affect the decision as to the employment of one oil in place of another with which it is interchangeable. It will thus be seen that the position of the crusher in normal times is not always easy, while that of the Government under present conditions is yet more difficult, when it has to decide what are the nuts or seeds of which the import shall be stimulated or restricted.

CONSUMPTION OF OILS IN DIFFERENT COUNTRIES

In estimating the general economic situation so far as oils and fats are concerned, the first consideration that presents itself is the amount of the various oils consumed by the principal countries. The table on p. 20, will show the pre-war importation of the various nuts and seeds. From this a tolerably accurate estimate can be obtained, if it is remembered that to the figures thus given home-grown supplies, *e.g.* of tallow and butter, must be added before any total can be considered as complete.

It will be seen that Germany was by far the largest consumer of these products taken as a whole, and that of palm kernels in particular she absorbed a quantity over three times as great as that imported by any other single country, the United Kingdom and the Netherlands coming next. As importers of copra France and the Netherlands followed Germany more closely, while the French importations of ground nuts amounted to nearly 500,000 tons, as against about 100,000 tons that were taken by Germany and 70,000 by the Netherlands. The importations of cotton seed into the United Kingdom far outstripped those of every other country, amounting to 615,000 tons as against 220,000 tons going to Germany. In the case of linseed the United Kingdom headed the list with an import of 650,000 tons, 560,000 tons going to Germany and considerable quantities to the Netherlands, Belgium, France and the

STATEMENT showing the IMPORTS for CONSUMPTION of certain OIL-SEEDS into the undermentioned Countries in 1913.

Imported into	Palm Kernels	Ground Nuts	Copra	Soya Beans	Cotton Seed	Linseed	Rape Seed	Sesame
	Met. Tons	Met. Tons	Met. Tons	Met. Tons	Met. Tons	Met. Tons	Met. Tons	Met. Tons
Germany	235,917	98,085	196,449	125,750*	219,797	560,323	153,427	116,039
France	2,986	493,467	112,640	45	17,670	237,406	61	20,586
Netherlands	63,711	67,428	100,635	27,554	(a)	286,035	47,908	(a)
Belgium	(a)	(a)	19,552	(a)	(a)	259,105	(a)	(a)
†Denmark	595	3,666	33,687	48,069	(a)	19,980	2,148	4,018
Sweden	(a)	9	(a)	(a)	(a)	28,876	914	(a)
†Norway	(a)	(a)	(a)	(a)	(a)	14,706	964	(a)
Russia	(a)	3,347	67,537	(a)	712	2,193	(a)	(a)
†Finland	(a)	(a)	(a)	(a)	(a)	2,726‡	(a)	(a)
	Tons of 2240 lbs.	Tons of 2240 lbs.	Tons of 2240 lbs.	Tons of 2240 lbs.	Tons of 2240 lbs.	Tons of 2240 lbs.	Tons of 2240 lbs.	Tons of 2240 lbs.
†United States	(a)	8,519	18,246	(a)	(a)	132,357	(a)	(a)
†United Kingdom	74,797 (in 1914).§	(a)	30,868	76,452	615,332	654,812	26,556	(a)

(a) Not separately distinguished.
* Including mowra seed, illipe and shea nuts and castor seed.
† Total imports for these countries.
‡ Including hemp seed.
§ Palm kernels were first shown separately in 1914.

United States. It was, however, an abnormal year as regards linseed, and the average importations over a series of years are not nearly so large. Finally, of the relatively less important products of rape and sesame seed, Germany imported much more considerable amounts than any other consumer.

It is, of course, evident that the whole position is not revealed by the above figures. The actual position of each consuming country can only be ascertained by converting the seeds into oil values and also by deducting the figures of re-export. Further, the figures are for a given pre-war year, and thus do not show either the abnormal conditions of the war, or the development in sources of supply.

So far as the United Kingdom is concerned, while the ordinary official statistics as to the import of certain nuts and seeds and manufactured oils were accessible, there was no accurate survey made as to the British consumption of each of the different oils before the war. For the years 1915 and 1916, however, calculations have been made by Mr. Pearson, the Chairman of the Oil and Seed Crushers' Association. They are based on the Board of Trade Returns, and show the quantities available for all ordinary trade requirements in each of those years. Even so, certain subheads (notably "Nuts and Kernels," and "Sundry Seeds") would well repay further sub-division, but the table compiled by Mr. Pearson has made the position much clearer than it was before. Similar calculations have now been made for 1912–14, and the average for those years is given in the table on p. 22, side by side with the figures for the subsequent years.

OIL: Average Figures for 1912–14 compared with those for 1915 and 1916.

	Oil: Imports and (or) Oil Equivalents of Seed Imports			Oil Exports			Oil: Net Amount Retained for United Kingdom.		
	Average 1912–14	1915	1916	Average 1912–14	1915	1916	Average 1912–14	1915	1916
	Tons	Tons	Tons	Tons	Tons	Tons	Tons	Tons	Tons
Linseed	158,124	123,554	151,710	27,562	55,927	24,761	130,562	67,627	126,949
Cotton-seed	127,510	117,911	60,212	26,360	36,082	3,397	101,150	81,829	56,815
Rapeseed	22,366	21,234	33,219	6,049	6,048	3,927	16,316	15,186	29,292
Sundry seeds (including soya)	38,933	87,751	72,142	16,878	16,565	6,197	22,055	71,186	65,945
Nuts—									
Copra, Palm-kernels, Ground-nuts	39,321	169,892	163,684	..	181	121	39,321	169,711	163,563
Coconut oil	52,012	49,663	28,869	9,381	14,095	7,285	42,621	35,568	21,584
Palm oil	79,244	67,381	62,178	47,383	40,874	18,546	31,861	26,507	43,632
Tallow	93,055	82,079	47,046	78,401	57,580	29,782	14,654	24,499	17,264
Stearine	6,251	6,576	2,857	409	1,553	194	5,842	5,023	2,663
Soap stock	3,567	1,448	309	..	..	..	3,567	1,448	309
Fish oil	63,811	91,872	109,061	8,229	4,570	1,146	55,582	87,302	107,915
	684,194	819,361	731,287	220,662	233,475	95,356	463,531	585,886	635,931
Animal oil	10,068	..	..	252	..	..	9,816	..	..
Lard	92,682	110,522	97,306	12,159	10,105	..	80,523	100,417	97,306
Imitation lard	11,596	10,717	..	308	52	..	11,288	10,665	..
Total	798,540	940,600	828,593	233,381	243,632	95,356	565,158	696,968	733,237

POSITION OF THE UNITED KINGDOM DURING THE WAR

During the war the oil position in this country has been affected both by the general considerations caused by the state of war and by special circumstances peculiar to the trade; and although these are not the main objects of this paper, it may be well to mention them briefly, in view of the fact that some are of more than an ephemeral nature and have an important bearing on the future.

The main factor in the situation has been the need for maintaining supplies here at a sufficient level in face of tonnage difficulties on the one hand, and the increased demand for oils and oil seeds on the other. This rise in demand was due to three main causes. Firstly, the requirements of glycerine for explosives involved splitting a quantity of oil far exceeding that normally so treated for the purposes of the soap trade. Secondly, oils are required for the manufacture of margarine to supplement the curtailed supplies of butter, and to prevent our becoming increasingly dependent on foreign countries for the substitute. Thirdly, it is important to maintain, so far as possible, the quantities of feeding cake available for stock, since any great deficiency must react on the supply of milk and home-grown meat, and also affect the fertility of the land through the absence of natural manure. An integral part of the same problem is the question of finding means to deal with the supplies once they are assured, and to convert them into the needed products. It is too soon to estimate yet how far these difficulties will be surmounted. All that can be said is that, up to the date on which this

CHART SHOWING PRODUCTION OF OIL AND FATS IN THE BRITISH EMPIRE.

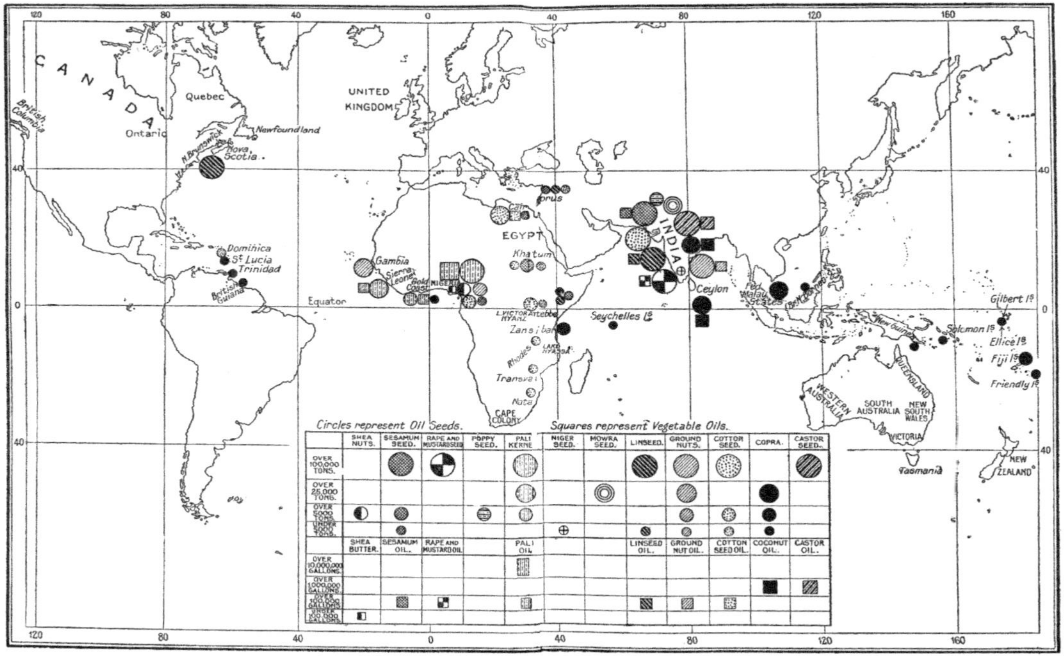

Reproduced by permission of the Admiralty Intelligence Department.

lecture was delivered, every effort had been made to deal with them comprehensively.

Fortunately, as against these difficulties, may be set the fact that there has been during the war no shortage of supplies of oils and fats, taken as a whole, in the countries of origin. It is true that a season's supply of one commodity from one region, such as tallow in Australia, or of linseed in Canada, may have fallen below the average, but the interchangeability of the various oils is an advantage from this point of view, and enables the gap thus created to be filled up from other sources. Further, the British Empire possesses the very substantial advantage that the supply of the various products produced within the Empire is amply sufficient for our own present requirements as consumers, and leaves a margin for export, if desirable. By the system of prohibiting exportation from the countries of origin to other destinations than the United Kingdom or the Allies, except under special licence, it has been possible so to organise and regulate this supply as to secure for this country any quantities required for which shipping was available. It has been further found possible to conclude arrangements with foreign countries to whom exports were thus licensed, under which they will send us an equivalent percentage of glycerine in return, thus in part solving the problem to which reference has been made above, of the abnormal demand for glycerine.

Position of Germany during the War

The figures of the pre-war importations of the different countries on p. 20 make it clear how large was the consumption of oils and fats in Germany. The stoppage of supplies to Germany sets free large quantities for use elsewhere. The neutral countries adjacent to Germany, however, have agreed to a system of rationing based on their pre-war consumption. The surplus, therefore, available for the Allies is proportionately increased, and its availability will depend entirely on the existence of adequate tonnage by which it can be brought. The chief effect of the blockade, however, consists in the very great deprivation which the curtailment of these imports has inflicted on Germany, both in view of the importance of fats for the maintenance of heat and energy, and of the fact that she was on the whole dependent for a larger proportion of these commodities from outside than of other classes of supplies. Indeed, it is probably true to say that the pressure of the blockade has told more severely on her in this respect than in any other. Unfortunately, the German propellant explosive is formed from nitro-cellulose and not nitro-glycerine. Otherwise the shortage of glycerine would have greatly accentuated the gravity of her situation.

Future Demand and Supply

So much for the position during the war. What is the outlook for the future? In the first place, apart from special demands due to the existence of hostilities, it is clear that the consumption of vegetable oils must be a rapidly increasing quantity. The advance of research and of industry is constantly opening up fresh avenues for their employment. But more than this:

it seems likely that instead of an increase of animal fats to meet the growth of population and of the demand consequent on a rising standard of life, there will be rather a decrease. In so far, therefore, as vegetable fats are interchangeable with them, the demand for the latter will be great, and the danger of any over-production in the world as a whole correspondingly small. From this point of view, therefore, it is clear that the sources of supply, actual or potential, are of great importance. A few of these may be mentioned. The consumption of copra in 1913 in the principal European countries and the United States amounted to a little under 600,000 tons. The principal sources of supply were the Dutch East Indies (200,000 tons), India (70,000 tons), the Philippines (70,000 tons), and Ceylon (30,000 tons). It must not, however, be imagined that these figures represent the total supply, that could be obtained at a price. Thus the actual crop, both in India and Ceylon, is much larger than the above figures. The islands of the Southern Pacific also represent a potential supply of which the development is yet only in its infancy. Of palm kernels about 375,000 tons were imported into the same countries in 1913, and of this total nine-tenths came from British West Africa. This supply is susceptible of large further increase, but it should be remembered that at least as much development may take place in the Belgian Congo and other foreign possessions as in British colonies. Ground nuts form a crop of much wider geographical distribution than palm kernels. Over 650,000 tons were imported in 1913, of which over 200,000 tons came from French West Africa, the same quantity from British India, and about 70,000 tons from British West Africa. The same remark, however, applies in this case as in that of other nuts. Thus, Indian exports only represented about one-quarter of the local crop, of which the greater portion was consumed locally. In the case of linseed the largest exporting countries are the Argentine Republic, India and Canada. The Canadian crop, however, besides being a diminishing quantity, is absorbed, as a rule, by the United States, only the surplus in bumper years finding its way to Europe. Argentine exports to Europe, on the other hand, in 1913 reached the abnormal total of over 1,100,000 tons, but a figure of 800,000 tons would not be unusual. Local consumption of the Indian crop again accounts for the larger part of the crop, but the exportable surplus varies from about 250,000 tons to over 400,000 (in 1913). Of cotton seed the largest crops are produced by the United States, India and Egypt. The American crop is almost entirely consumed at home, what export there is being of cotton-seed oil or cotton cake. The Egyptian export of cotton seed in normal years may be placed at about 300,000 tons. In the case of India the local production is very large, reaching 2,000,000 tons. The quantity exported, however, chiefly to the United Kingdom, varies from 100,000 to 400,000 tons, a slight difference in price being sufficient to bring forward largely increased quantities. Soya beans are a product of the Far East, China, Manchuria and Japan. But their popularity in Europe has decreased, and imports, therefore, have diminished from over 400,000 tons in 1910 to a much lower figure.

The Economic Position of the British Empire

The foregoing figures deal only with the best known oil seeds, nuts and beans. They are sufficient, however, to show what a large part of the exportable surplus of the world is grown within the British Empire. This conclusion is strengthened if it be remembered that of other seeds, such as rape, sesame and mowrah, India is by far the largest producing country. The whole position can, indeed, be succinctly gathered from the annexed map. What, then, are the economic deductions that can be drawn? In the present time of war the advantage of the British position is obvious, so far as supplies are concerned, and it is effectively exploited by the system of prohibitions and licences, which has been already explained. But when peace returns the question will also be one of great importance. The existing production in India is very great, and no doubt is capable of still further increase. But the latent potentialities of other British possessions are enormous, especially as regards the most valuable of oils, coconut oil, palm-kernel oil and ground-nut oil. Clearly, therefore, in view of the increasing requirements of the Western world, they should be developed. The further suggestion, however, has been made that the British Empire possesses in its store, actual or potential, of such oils, a great monopoly which it has been able to exploit in war, and of which it should be able to make an analogous use in peace time. But before any action can safely be taken on such a hypothesis, it needs to be analysed with considerable care. Palm oil has, indeed, for certain purposes almost a monopoly value. But in the case of other oils the subject needs to be approached with great caution. It is true that the demand will probably for some years outrun the supply. But it must not be forgotten that many other potential sources of supply are capable of development in other countries than the British Empire, *e. g.* in the case of copra. Not only so, but either fresh discoveries or the introduction of new processes may make available oil products which have not hitherto been utilised. And, lastly, the constant progress of chemical research tends to make applicable to higher uses oils which hitherto have been employed only for commoner purposes. While, therefore, a measure like the recently proposed export duty on palm kernels is economically justifiable, the principle cannot be pushed too far. Be, however, that as it may, the potentialities are enormous which the British Empire possesses in the strength that comes from the development by ordinary peaceful trading of a rich inheritance, while the same resources have been of incalculable advantage in time of war.

THE SUGAR SUPPLY OF THE EMPIRE, ITS PRODUCTION AND DISTRIBUTION[1]

By C. SANDBACH PARKER, M.A., C.B.E.

SUGAR has from time immemorial been a political and fiscal question in this country, bristling with difficulties and fraught with disaster to most of those who attempted to legislate for it, or depended for their living on growing it under the British flag. The past history is mainly bound up with the West Indies, which were at one time practically the sole British source of supply for this country. From 1660 up to 1854 British colonial grown sugar was admitted on preferential terms into this country. From 1787 onwards the import duty on British sugar was 29*s*. per cwt. = over 3*d*. per lb. That on foreign sugar 45*s*. 6*d*. = nearly 5*d* per lb. In 1841 Cobden led the agitation in favour of Free Trade. There followed an

[1] In preparing my paper I have received great assistance from the various papers prepared by Mr. George Martineau, C.B., who has devoted himself for the last sixty years to the study of the subject and who speaks with unsurpassed knowledge. His latest contribution entitled " A Short History of Sugar, 1856–1916," written for and published by the British Empire Producers Organization, of which I have the honour to be chairman, is perhaps the most valuable of his contributions from the point of view of the general public, who can in its few short pages gather a comprehensive and complete view of the circumstances leading up to the present scarcity of sugar in this country. I have also to thank Messrs. Czarnikow and the *International Sugar Journal* for valuable assistance.

eight-day discussion in the House of Commons on the Sugar Duty Bill of Lord Melbourne's Government, which was rejected, and a motion of want of confidence by Sir Robert Peel was carried by one vote. Parliament was dissolved, the Government was beaten, and Sir Robert Peel became Prime Minister. In 1842 he carried a revision of the Customs Tariff and the introduction of an Income Tax. In 1845 the Income Tax was renewed, and the sugar and other duties reduced. In 1854 differential sugar duties were finally abolished, and slave-grown sugar from Cuba and Brazil admitted on the same terms as British colonial sugar.

In this connection I give the following extract from an article I wrote for the *Fortnightly Review* in 1898 entitled "Free Trade and Cheap Sugar"—

"The following extract from a letter, written by Mr. Goulburn to Sir Robert Peel in 1847, referring to the admission of slave-grown sugar into competition with our Colonial sugar under the same scale of duty, will show the views of a former Chancellor of the Exchequer on Free Trade. He writes—

> "'I deeply regret having, on political grounds, quite independent of colonial interests, voted for the equalization of the duties on British and Foreign sugar. It was in itself wrong. It applied to the produce of the West Indies what was not applied to the produce of other British Colonies, and, so far from carrying out a principle of Free Trade, was in direct violation of it, so long as the West Indies were subject to limitations as to labour from which foreign countries were exempt.'

"He goes on to suggest the repeal of the Act abolishing differential duties, the effect of which would be 'to restore confidence and to induce those who are now disheartened to go on.' This is a capital illustration of the present position. The abolition of differential duties on slave-grown sugar allowed foreign interference in our home markets by the admission of sugar receiving a bounty in the shape of slave labour into competition with sugar grown in our own Colonies where we had forbidden slave labour. By the efforts of our Government this grievance has now been removed by the abolition of slavery. But it has been replaced by a more direct form of foreign interference, *viz.* bounties on export, which have gone on increasing ever since their inception, stimulated by the existence of a so-called Free Trade country at their very doors, available as a dumping ground for as much bounty-fed sugar as those countries cared to send."

The Memoranda and Statistical Charts of British and Foreign Trade and Industrial Conditions, published in 1903,[1] makes the following statement—

"The equalization of duties on Raw Sugar and the graduation of those on Refined (1846–1854) was soon followed by the introduction of foreign Beet Sugar, both raw and refined from Europe."

At the beginning of the nineteenth century, this country drew practically the whole of its supplies from the West Indies, which were in a flourishing condition. The production of sugar in the British West Indian Islands, excluding British Guiana, in 1802 was over 170,000 tons. In 1913–14 it was about 134,000 tons. The British Guiana crop in 1913–14 was 100,000 tons. The prosperity of the sugar producing industry in the West Indies reached its zenith in the early part of the nineteenth century, but the industry was reduced to the verge of ruin by the end of that century, from the following causes—

(*a*) The abolition of slavery in 1832 and the inadequate compensation of the planters.

(*b*) The reduction and abolition of the preferential duty in this country, 1846–54, which exposed them to competition from slave-grown sugar in Cuba and Brazil.

[1] C.D. 1761.

(*c*) The growth of the European beet industry fostered by bounties, cartels, freight subsidies, etc., which practically excluded British West Indian sugar from the market in this country.

I now propose to sketch the development of the European beet industry. The fact that sugar could be obtained from beetroot was discovered by a German chemist, Marggraff, in 1747. In 1800 the King of Prussia started beet-growing in Silesia. During the Napoleonic wars the blockade of France by Great Britain prevented her obtaining sugar from her West Indian colonies, and led to the further investigation of possibilities of beet-sugar cultivation in France, and the establishment of the industry on an extensive scale in 1812–13. In 1853 the world's production of sugar was under 1,500,000 tons, made up of

European Beet	14 per cent.	=	200,000 tons.
British Colonial Cane . .	17 ,,	=	255,000 ,,
Foreign Cane	69 ,,	=	1,035,000 ,,

The bounty system commenced almost imperceptibly about 1860 by means of drawbacks on beet sugar exported. The growth of the production of beet sugar due to this assistance, coupled with the open market in the United Kingdom, is best shown by the following figures showing German progress only—

1871–2 . . .	Beet Crop . . .	186,442 tons.
1884 . . .	,, . . .	1,123,030 ,,
1913–4 . . .	,, . . .	2,738,000 ,,

The German method and foresight is well illustrated by the period 1871–84, when they levied their duty upon an assumed yield of sugar from the roots, and allowed drawback on the actual sugar obtained and exported, thus offering a strong inducement to farmers to improve the quality of their roots. The result was, that the sugar contents of the roots was increased in that period from 8·28 per cent. to 11 per cent., and the present average yield is about 17 per cent.

France during the same period levied the duty and granted the drawback on actual sugar, thereby offering no inducement to better agriculture; and in 1884 she was still only obtaining about 6 per cent. as against 11 per cent. in Germany from the roots. The production of France in 1884—272,962 tons—was slightly lower than in 1871–2—287,444 tons. In 1884 the French adopted the German system, with immediate results, their yield jumping the next year to 7·83 per cent., and by 1897–8 they had reached 11·40 per cent. against the German 12·79 per cent. About 1891–2, Germany, having achieved a high standard in sugar agriculture and manufacture, abandoned her system of indirect bounty and adopted the French system of levying the duty on the actual sugar produced, and in 1892 she adopted a system of direct bounties on export. In 1896, the French crop having risen to 668,516 tons, and the German crop to 1,821,233 tons, Germany doubled her bounty as a blow at the French industry, and this was followed by the granting of an equivalent export bounty by France in addition to the indirect bounty.

Meanwhile, what was happening in the West Indies and Great Britain? In 1852–61 the average imports to the United Kingdom were 436,000 tons, viz.—

	Tons
British Colonial Cane	275,000
Foreign ,, ,, ...	129,000
Beet	24,000
Unenumerated	8,000

In 1869–70 the world's production had grown to 2,600,000 tons, of which beet comprised 32·7 per cent. and cane 67·3 per cent.

In 1871 the consumption in the United Kingdom was about 700,000 tons, of which about 610,000 tons was cane refined in the United Kingdom and 90,000 tons refined imported from abroad.

In 1897 the consumption in the United Kingdom was 1,441,986 tons, made up as follows—

	Tons	
Foreign Refined	791,600	= 54 per cent.
,, Raw Beet . . .	476,016	
,, Cane	250,595	
	1,518,211	Total Imports.
English Refined	602,000	= 40 per cent.

Prices had fallen in the United Kingdom for—

Refined sugar from 37*s*. per cwt., excluding duty, in 1871 to 12*s*. 3*d*. in 1897;

Unrefined sugar from 25*s*. 10*d*. per cwt. in 1871 to cane 9*s*. 8*d*. and beet 8*s*. 11*d*. in 1897.

				Tons			Tons
In 1871–2	Germany	exported	Refined	5,809		Raw	Nil
,, 1897–8	,,	,,	,,	503,928		,,	478,941
,, 1871–2	France	,,	,,	98,345		,,	Nil
,, 1897–8	,,	,,	,,	146,716		,,	306,418
,, 1871–2	Austria	,,	,,	Nil		,,	Nil
,, 1897–8	,,	,,	,,	460,154	1896–7	,,	101,485 [1]
					1897–8	,,	31,449 [1]

[1] Due to higher drawback on Refined.

I do not attempt to deal with the amount of the bounty and cartel bounty in each country, which varied and would require lengthy explanation. In fact, the intricate calculations for bounties and cartels were hardly understood by our officials. But the effect of the bounty system was not only to stimulate good agriculture, selection of roots and efficient manufacture, but to increase the exports to the extent above named, while its effect on British production, which had no assistance from Government, was to establish practically a monopoly for beet in the markets of the United Kingdom. During this period—1854–97—nothing whatever was done to stimulate production in the West Indies, nor to counteract the adverse effect of the bounties on British industry. The market in the United Kingdom was practically closed to their product.

In 1884 the United States desired to obtain sugar from the British West Indies, and satisfactory reciprocal arrangements had been made when Great Britain stepped in and said that no arrangement could be ratified, as it was contrary to the "most favoured nation" clause in our commercial treaties. From 1884 to 1897 the British West Indian industry struggled on against constantly increasing competition from beet, which by 1896–7 had increased to 63 per cent. of the world's production. In 1897 the United States imposed countervailing duties on German sugar equal to the amount of the export bounty plus the cartel or indirect bounty. Canada began to give preferential treatment to West Indian sugar in 1898, by a 25 per cent. rebate of duty (= 18 cents per 100 lbs.), but the United States' countervailing duty of 27 cents

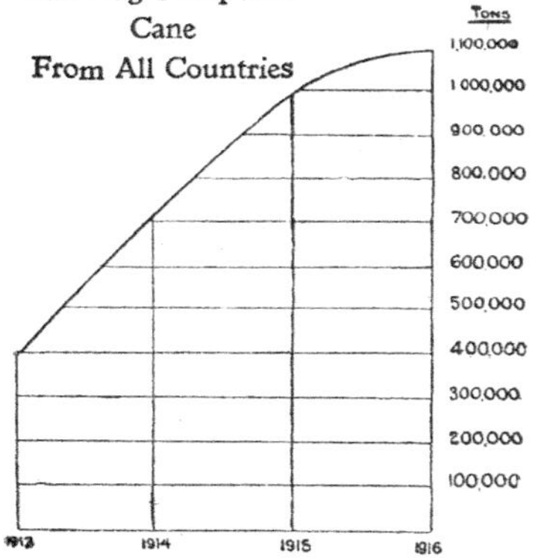

Raw Sugar Imported
Cane
From All Countries

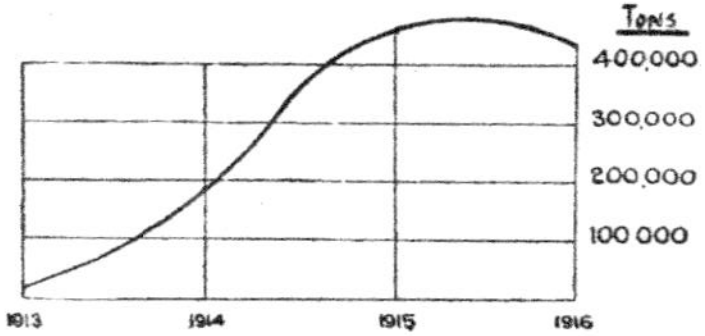

Refined Sugar Imported
Cane
From All Countries

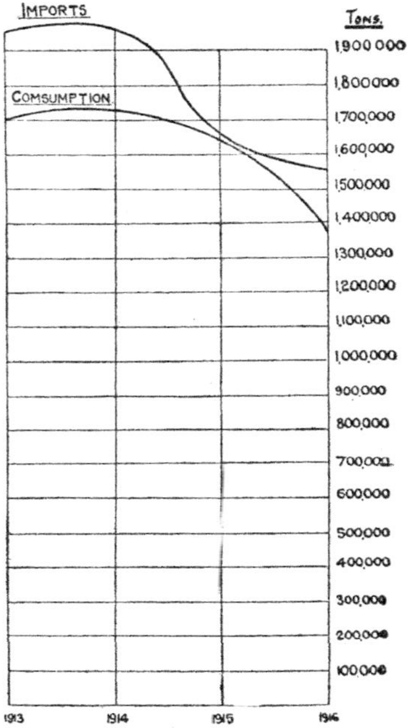

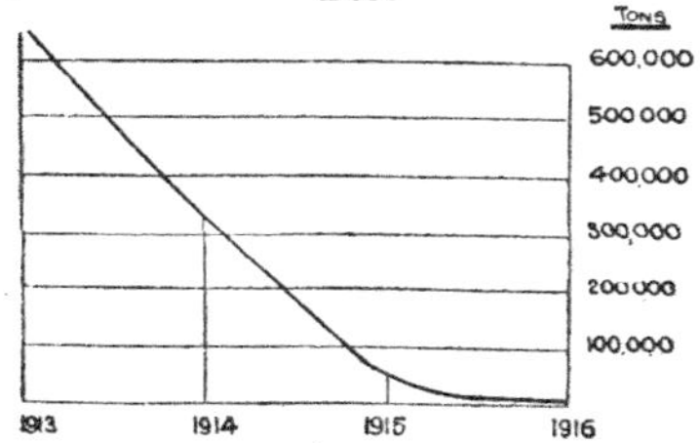

Raw Sugar Imported
Beet

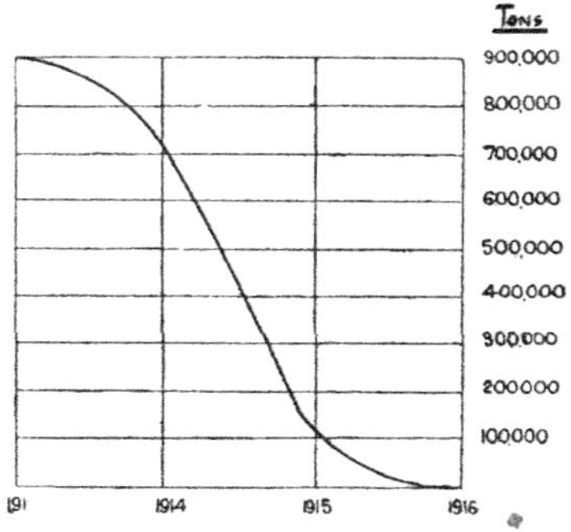

Refined Sugar Imported
Beet

against beet made it still the best market for sugar from the British West Indies.

A Royal Commission was appointed in 1897 to inquire into the state of the West Indies, and reported in the same year, indicating impending ruin to the sugar industry in those colonies. In 1898 Mr. Chamberlain carried his Bill for their relief, whereby a sum of £250,000 was granted to save them from ruin.

Direct bounties on export were abolished by the Brussels Convention in 1902, and the countervailing duty in the United States then ceased. The States had been the best market for British West Indian sugar whilst countervailing duties lasted, but thereafter Canada became the only market, increasing their rebate in 1900 to 33⅓ per cent. (= 24 cents per 100 lbs.). The beet industry in Europe was too firmly established to be seriously affected by the abolition of bounties. Central European producers retained possession of their home markets by reason of a prohibitive Customs duty; sugar was cheaper than ever, general prosperity greater, and the home consumption, sold at world's price plus duty, increased rapidly. The European beet crops reached their maximum, 8,341,063 tons, in 1912–13, but the cane crops, which were in 1902–3, excluding India, 3,628,000 tons, increased to 6,668,000 tons, excluding India, in 1912–13, and for 1916–17 are estimated to be about 9,000,000 tons, excluding India, partly as the result of the Brussels Convention, but mainly from the development in the colonies of the United States and Cuba, under the stimulus of preferential treatment in the United States.

Canada in 1903 imposed a surtax on German sugar of 33⅓ per cent. of the duty, which practically put an end to imports from Germany, who retaliated by placing a surtax on imports from Canada. These were removed in 1910, Germany coming to the conclusion that they and not Canada were the losers by the retaliatory surtax, and the British preference in Canada being sufficient to keep out German sugar. Had the Canadian market been capable of absorbing a largely increased crop of sugar, and had the preference accorded to the British West Indies been firmly secured for any term of years, an immense stimulus would have been given to the industry there. But this was not so. Canada's consumption in 1899 was only 118,671 tons; it was not until 1908 that her consumption rose to 208,709 tons, about the exportable production of the British West Indies, excluding grocery sugar sent to the United Kingdom. In 1909 the Canadian Government granted to their refiners the privilege of importing 20 per cent. of their requirements in foreign sugar at the British preferential tariff rate, and this to a large extent nullified the benefit to the British West Indies of the preference. In 1912 the Canadian West Indian trade arrangement abolished this privilege, and secured to British sugar for a period of ten years a preference of 20 per cent. of the tariff, not to be less than 15 cents per 100 lbs. for 96° sugar, say 13*s.* 4*d.* per ton. This arrangement was the direct result of the Royal Commission on Trade Relations between Canada and the West Indies, presided over by Lord Balfour of Burleigh.

It will be clear from these facts that until 1912

there was no certainty that the Canadian preference would be continued, and therefore no inducement to the British West Indies to materially increase their output. Meanwhile the policy of the United States to her colonies and dependencies had resulted in an immense development of cane production in those countries and of domestic beet.

		Tons	
Hawaii Crop	1898	204,000	Admission free.
	1916–17	575,000	
Porto Rico	1898–9	54,000	"
	1916–17	450,000	
Philippines [1]	1898	180,820	Of which 27,997 to U.S.A.
	1911	207,219	Of which 196,000 to U.S.A.
	1915–16	316,450	Admission free to U.S.A. since 1914.
Cuba	1903	1,003,873	20 per cent. preference since 1903.
(Estimated)	1916–17	3,400,000	
U.S.A. Beet	1898–9	32,471	Free.
	1916–17	775,000	

From 1897 to 1913 the duty on 96° sugar in the United States was \$1.68½ per 100 lbs., *i. e.* about £7 17*s.* 4*d.* per ton, and the Cuban duty \$1.348 per 100 lbs. (= £6 5*s.* 10*d.*). In 1913 the general tariff was reduced to \$1.256 (= £5 17*s.* 4*d.*) per ton, and the Cuban duty to \$1.0048 per 100 lbs. (=£4 13*s.* 9*d.* per ton). Cane sugar in the colonies of the United States having free entry has increased by over 900,000 tons, or 200 per cent., in nineteen years, while Cuban preferential sugar has increased by about 2,400,000 tons (= 240 per cent.) in fourteen years. The Cuban

[1] A more rapid expansion of cultivation in the Philippine Islands was prevented by legislation which gave the natives certain rights, and discouraged the investment of capital from countries like England and France.

crop has increased by about 800,000 tons since war broke out. The estimated crop of 3,400,000 tons for 1917 has been much reduced by the rebellion and by poor juice. Domestic beet has increased by 742,529 tons = 2284 per cent. in nineteen years.

The total American, domestic, colonial and Cuban crops have increased in nineteen years from 1,035,538 tons to 5,676,250 tons estimated for 1916–17 (in the last five years by about 1,500,000 tons), or nearly 2,000,000 tons in excess of American consumption in 1916. The total crops of the British Empire, excluding India, which has oscillated between 1,900,000 and 2,600,000 tons, have only increased in the last nineteen years by 467,000 tons to 1,015,000 tons estimated 1916–17, and are still short of the Empire's requirements by about 2,500,000 tons. I have dealt with these figures in detail in order to satisfy you on the following points—

(1) Had the reciprocity treaty between the British West Indies and the United States been allowed to go through in 1884, similar results to those attained in Cuba might have been achieved in our colonies.

(2) Lack of security that the Canadian preference would continue, and the small requirements of that country did not justify any extension in the West Indian industry, though it opened a market for its sugar at a time when no other market was available.

(3) The opening in the United States in 1897 for British West Indian sugar due to countervailing duties could not have been of long duration, owing to the rapid development of the American new possessions from 1898 onwards.

(4) Had England extended similar treatment to British-grown sugar, a rapid development would have taken place in the British West Indies and all other cane-growing countries of the Empire. The savings to the nation in this war would have been colossal. The imports of foreign sugar into the United Kingdom in 1916 were approximately—

	Tons
Java	316,834
Cuba	550,206
American Granulated	268,465
Other countries outside the British Empire	138,500
Total	1,274,005

This import at a cost of, say, £20 per ton, means about £25,500,000 paid to countries which do not import one-tenth of this value from the United Kingdom in war time. The drain of gold in three years' war will be little short of £80,000,000, which, had the Government policy been less short-sighted in the past, might have been kept in the British Empire.

Let us look into this matter more closely, and try to follow whither the golden sovereigns flow, filter through, that we paid, say, for Java sugar. The sugar factories in that island are owned by private planters, by Dutch corporations, or even by banks. The payments are made in Java, and our money is partly used for manure and machinery. Where the manure is bought is an open question, but most of the machinery comes from Holland, some of it came formerly from Germany, and a little from Great Britain. The large surplus of profits above the expenses is either used for enlargements, increase of production in Java, or invested at home, in Holland. How is it invested there? Partly, no doubt, in State securities, but a greater part, probably, in bank shares, shipping shares, shipbuilding, and in shares to increase the productions of condensed milk or margarine. Where are these products sent to? Is it not possible that some of our money finds its way into German securities or German industries?

It has frequently been stated on hearsay, and is commonly believed, that the industry in the British West Indies was brought to the verge of ruin by neglect and lack of initiative on the part of those engaged in it. The above facts show that, so far from this being the case, they made a magnificent fight against heavy odds in the bounty-assisted beet industry, and were only saved from complete ruin by the abolition of bounties and the Canadian preference.

The Refining Industry in the United Kingdom.—Whilst refining in all other countries progressed with the increasing requirements of their markets, in England the reverse was the case. In 1860 we refined all our requirements. In 1913 we refined less than half. Competition from bounty-fed refined sugar, from Germany and Austria chiefly, ruined the industry, which became concentrated mainly in three places—London, Liverpool and on the Clyde. Our refining capacity at the outbreak of war was about 850,000 tons per annum, but there is no reason why, with proper security, we should not refine all our own requirements.

Beetroot Industry in the United Kingdom.—Two or

three attempts have been made to establish this in England. The first was in the 'eighties, when a factory was established at Lavenham in Suffolk, but it was obviously impossible to hope for success against the German bounty-fed industry, and it soon failed. More recently, in 1912, another attempt was made in Norfolk, where the Cantley factory was erected, but this has not been a success owing to the insufficiency of beets to keep the factory running at an economic rate; and on the refusal of the Dutch Government to allow the export of seed it had to close operations. Various causes contributed to its want of success, but in my view one of the main reasons was that the company did not itself own beet land, but depended on contracts with farmers for their supply of roots. Farmers in this country are a conservative race and slow to experiment with new crops, and their treatment by our Government was not of a nature to encourage any new departure. In 1915, 20,000 tons of roots only were obtainable, against a requirement of 60,000 tons, but the sugar contents averaged 17·36 per cent., which compares very favourably with continental results, and justifies the contention that beet can be grown as well in the United Kingdom as in Germany. All that is wanted to establish the industry is preferential treatment, and the ownership by the factory companies of sufficient land in their immediate neighbourhood to secure to each factory sufficient beets to enable it to be run at an economic rate.

The beneficial effects of beetroot cultivation on the succeeding crops in the rotation are not yet realized by farmers in England, nor the value of the residue for stock-feeding purposes.

The Report of the Agricultural Sub-Committee of the Reconstruction Committee, appointed by the Government in August 1916 to investigate after-war questions of industry and economics, includes the following statement with regard to the introduction of the beet industry into England—

> "We believe that the advantages to be derived from its successful establishment would be very important. . . . The nation by producing a proportion of its own requirements of sugar would reduce its dependence on imported supplies. . . . The rural community would be benefited and augmented by the provision of employment in sugar factories in the winter months, and of additional work on the land in the summer. Agriculture would gain by the introduction of a new crop, which yields a satisfactory cash return to the farmer and leaves a residue which is a valuable cattle food. Moreover, wherever beet was grown, the deep tillage and improved cultivation necessary for its success would prove of immense benefit to the other crops in the rotation as well as an object lesson to the country in good farming. Of all these statements, there is abundant evidence from Continental sources."

A fresh attempt to establish the industry has recently been announced. The British Sugar Beet Growers' Society has just purchased an estate of 5600 acres at Kelham in Nottinghamshire, and intends erecting a factory capable of dealing with 1000 tons of beets *per diem*. The possession and cultivation of so large an area by the society should, if the necessary security is given, by legislation ensure the success of the venture.

The production of sugar in other parts of the British Empire, with the exception of Mauritius, is consumed locally.

INDIA produces about 2,400,000 tons and has to import about 800,000 tons. This used to come largely from Germany and Austria, but the Indian Government put a stop to this by imposing countervailing duties equivalent to the bounty in 1899. In 1902 India imposed an extra import duty equivalent to the cartel profit. These duties were abolished after the Brussels Convention in 1903, and since then the influence of bounties having been removed, India's import has been chiefly from Java and Mauritius. The industry is capable of great development, and the Government is stated to have thrown open 60,000 acres in the Bombay Presidency for sugar cultivation, with an irrigation scheme attached.

NATAL produced about 92,000 tons in 1913–14, and is estimated to produce 120,000 tons in 1916–17. The industry is capable of great development, but the crop is at present limited by the consuming capacity of South Africa. Preferential treatment in the United Kingdom would greatly stimulate production.

AUSTRALIAN production is in a serious position owing to labour troubles. The crop is about 220,000 tons, but factory extensions capable of bringing up the crop to 300,000 tons are stated to have been recently made.

FIJI production is about 80,000 tons, and is practically absorbed by Australia and New Zealand. A small proportion goes to Vancouver.

MAURITIUS produces about 250,000 tons, and is not capable of much development.

Practically none of the sugar produced in the Empire had, prior to the war, come to the United Kingdom, except from the British West Indies.

SUGAR DURING THE WAR PERIOD

This brings us down to the outbreak of war in August 1914, which found this country in a state of complete dependence upon enemy powers for two-thirds of its sugar supplies, a refining industry capable of dealing with less than half the requirements, and the Overseas Empire production barely equal to the consumption of the Empire consuming markets outside the United Kingdom, excluding India.

When war broke out in 1914 our annual consumption was over 1,700,000 tons, and our imports over 1,900,000 tons, of which only 71,661 tons came from British sources; and about 1,300,000 tons from Germany and Austria, of which 663,517 tons was refined sugar.

The balance, 600,000 tons, came from other foreign countries. Of this total about 400,000 tons went to the sugar-using trades, 100,000 tons to brewing trades, and the rest (equal to about 58 lbs. per head of population per annum) was directly consumed. This means a weekly allowance of 1·12 lb. per head against the present ¾ lb. allowed, since reduced to ½ lb.

The Government found themselves face to face with an unprecedented famine in this country, as 1,300,000 tons was immediately cut off. They took prompt steps to remedy this. The Royal Commission on Sugar Supplies was appointed and proceeded to buy something like 1,000,000 tons of sugar. The world's price, which on August 1, 1914 (New York quotation for 96° crystals excluding duty), was about 9*s*. 6*d*. per cwt. c. & f.

E

It rose in three weeks to 24*s*. 7*d*. c. & f., and was at the end of the month 21*s*. 10*d*. c. & f. The Government have been severely criticized at times for the policy pursued at the commencement of the war in making such large purchases and driving up the world's market price. Their purchases during the first two months of the war amounted to 1,288,806 tons, or almost the exact equivalent of a year's imports from Germany and Austria. They prohibited the export of sugar from British colonies, in order to secure as much as possible for this market, which for the past thirty years had been practically closed to those sugars. This was a most serious blow to British Guiana, whose heavy crop season was about to commence. Sugar had been sold and was being offered for sale to Canada. I was deputed with Mr. Campbell, Chairman of the West India Committee, to negotiate a sale to the British Government represented by Mr. McKenna, the then Chancellor of the Exchequer. We sold him some 56,000 tons at 17*s*. per cwt. f. o. b. I pointed out at the time that the price was far below the value of the sugar, and it is interesting to note that about 100,000 tons of Javas were purchased practically at the same date, August 24, 1914, at 17*s*. 6*d*. to 17*s*. 9*d*. f. o. b., against 17*s*. f. o. b. paid for British Guiana sugar, which was at the date of sale worth about 21*s*. 8*d*. f. o. b. for Canada, at which price similar sugar was sold by my company within a fortnight.

It has always been a satisfaction to the British West Indies to feel that after they had been persistently neglected and brought to the verge of ruin by the policy of the Mother Country, they were called upon to save her from famine and did so at a loss of something like £250,000 to British Guiana alone, in the price at which they sold their sugar to the Government.

The Royal Commission on Sugar Supplies continues to provide for and control practically the whole consumption of sugar in this country.

The following figures from the Board of Trade returns, per W. Connal's circular, show the consumption of the last four years, 1913–1916 (December 1 to November 30)—

	Imports		Refined		Total Imports	Consumption
	Raw Cane	Raw Beet	Cane	Beet		
	Tons	Tons	Tons	Tons	Tons	Tons
1913	395,672	645,970	670	905,952	1,948,264	1,719,976
1914	714,224	331,991	187,346	710,100	1,943,661	1,729,243
1915	1,074,467	4,273	464,148	103,264	1,646,152	1,650,551
1916	1,118,758	—	423,500	9,587	1,551,845	1,395,569
	3,303,121	982,234	1,075,664	1,728,903	7,089,922	6,495,339

The total purchases by the Sugar Commission to the end of 1915 were—

	Tons
1914	1,288,206
1915	1,040,690
	2,328,896

The figures for 1916 have not been published.

It is interesting to note that the production of the British Empire has increased since the war by nearly 400,000 tons, viz.—

			Tons
Increase—			
British West Indies and British Guiana			86,000
Egypt			55,000
Natal			28,000
India			300,000
			469,000
Decrease—			
Mauritius	30,000		
Australia and Fiji	50,000		80,000
		Net Increase	389,000

The reductions are temporary.

The distribution of sugar has latterly given rise to considerable discontent in this country. For months past complaints have been made that the poor in many places cannot get sufficient sugar for their day to day requirements owing to grocers insisting that they should, as a condition of getting any sugar, buy a certain value of other goods which in many cases they cannot afford. This feeling of discontent is aggravated by the fact that there seems to be no scarcity of sweets anywhere. It is hard to understand why this difficulty still exists, as the ration of ¾ lb. per head per week should not absorb more than 800,000 tons, and the reduced supplies to the sugar-using and brewing trades (40 per cent. for the former and 25 per cent. for the latter) should absorb under 200,000 tons, in addition to which there is the quantity consumed by the Navy and Army, which is estimated at not exceeding 100,000 tons. Total 1,100,000 tons out of over 1,500,000 tons imported in 1916. But unless better distribution results in enabling all to get their allowance of ¾ lb. per head per week, it would appear in the public interest that the quantity allocated to the sugar-using trades should be further reduced. The sugars used by the brewing trade are unsuitable for ordinary consumption, and are of secondary value for refining purposes.

The problem of future supplies during the war is mainly one of tonnage available to carry sugar to this country. The question of price remains to be dealt with.

Messrs. Willett & Gray give some interesting statistics of the retail price of granulated in various countries in 1911 as follows—

	Cents			Cents	d.	Consumed per head of population lbs.
U.K.	5	= 2½d. per lb.	Duty	·40	= ·20	85·38
U.S.A.	5·7	= 2·85 ,,	,,	1·34	= ·67	82·43
Germany	5·9	= 2·95 ,,	,,	2·03	= 1·015	41·27
Austria	6·5	= 3·25 ,,	,,	4·02	= 2·01	24·77
Russia	7·2	= 3·6 ,,	,,	8·56	= 4·28	22·77

"While the working men in Europe average to earn but one-third to one-half the wages earned by American working men, they average to pay 45 per cent. more for a pound of second or third grade granulated sugar than American workmen pay for the purest and handsomest sugar in the world."

Daily wage rates of agricultural labourers in beet fields—

U.S.A.	$2.60	= 10*s.* 10*d.*	all kinds of labour.
Germany	·47	= 1*s.* 11·5*d.*	men only.
Austria	·30·2	= 1*s.* 3·1*d.*	,, ,,
Russia	·24·9	= 1*s.* 0·45*d.*	,, ,,

In Factories

U.S.A.	$2.29	= 12*s.* 5·5*d.*
Germany	84	= 3*s.* 6*d.*

Although we were told that the Germans were being charged prohibitive prices for sugar in order that it might be exported cheap to this country, we find the consumption increased faster than our own, as shown by the following comparison—

	1888–9 lbs.	1914–15 lbs.
Germany per head	17·70	74·95
U.S.A.	50·44	86·04
Austria	13·01	37·38
U.K.	73·24	89·69
Russia	10·16	29·26

Revenue derived, Internal and Customs, from sugar, 1912–13—

	$		$	Per head of population
Germany	43,255,548.00 =	£ 9,011,572 =	0.666	= 2*s*. 9·3*d*.
U.S.A.	53,134,244.00 =	11,069,634 =	.59	= 2*s*. 5·25*d*.
Russia	77,028,550.00 =	16,047,615 =	.517	= 2*s*. 1·85*d*.
U.K. (1913)	14,851,517.00 =	3,094,066 =	.333	= 1*s*. 4·6*d*.

Sugar which before the war cost the consumer in the United Kingdom about 2*d*. per lb. now costs over 5½*d*. The increase in duty is responsible for 1·3*d*. per lb. The rest is due to the rise in cost of actual sugar due to the shutting off of pre-war supplies, and it must not be forgotten that the Royal Commission has an immense advantage in the freights on imported sugar; but for this, the price to consumers would be higher still. In those parts of the Empire where sugar is produced for their own consumption, there has been no corresponding rise in price.

In U.S.A. the average price of granulated in 1912 was 5·041 cents (= 2½*d*.) per lb. = 3·141 cents bond.

In 1916—6·862 (= 3·93*d*.) per lb. = 5.50 bond.

The reduction in U.S.A. duty since 1912 is about half a cent per lb. refined.

In U.K. the price in 1913 was *2d.* = 1·8*d.* bond.

In U.K. the price in 1916 was 5½*d.* = 4·0*d.* bond.

In Australia and New Zealand the price remains about normal, 3½*d.* and 2¼*d.*

It is clear that our dependence upon Germany and Austria had two effects on the outbreak of war—

1. It caused an enormous rise in price here, which need never have taken place had the Empire produced sufficient sugar for its own requirements.

2. It caused a big rise in price in the United States because we had to draw on Cuban supplies to fill the gap.

The effect of the high prices so caused has been an immense development of production in Cuba and Java, where, owing to immunity from excess profits and income tax, the sugar producers have made stupendous fortunes at the expense of the consumers, chiefly in this country. It has caused little or no development in British dominions, because those producers who reside, or whose companies are registered, in this country have to pay the bulk of their extra profits away to the Government in excess profits and income tax. Furthermore, the staffs of British sugar estates have been depleted to fill army requirements. New machinery and necessary repairs have been difficult, if not impossible to obtain.

The recent prohibition of the import of rum and the export of sulphate of ammonia, and inability to obtain other necessary fertilizers, all indicate that at the end of the war the British sugar industry will, unless some change in the policy of this country is made, be absolutely crushed out except in those parts of the Empire where it is protected.

Sugar after the War

I have indicated above the position of the British sugar industry when the war comes to an end, and I will now deal with after-the-war conditions.

1. It is nearly certain that no return to pre-war values can be expected, at all events for a considerable period after the war. The shortage of labour and other causes make it highly improbable that Germany or Austria will have any sugar to export, and, in fact, they may require to import. The devastation of the best beet-growing districts in France, Belgium and Poland indicate that these countries will be buyers. Consumption in this country will at once increase with the removal of restrictions.

2. The surplus production of U.S.A. colonies and Cuba will to some extent fulfil the immediate requirements and will control the price, but more will be required—we must either (*a*) resume imports of beet sugar from Europe, or (*b*) the British Empire must develop its own production. Should the former be necessary we should look to Russia and France for our supplies.

3. No development will take place in the British Empire unless security of market can be assured for a reasonable term of years.

4. Security in this, the only large market in the Empire, except India, which does not control its own

supply, can no longer be obtained on the same terms as before the war.

Conditions have completely changed. The value of money has risen, the cost of everything has risen, and capital cannot be tempted into development of the sugar or any other industry on the same terms as before the war.

5. The enormous fortunes made in Cuba and Java during the war have largely been spent, in the former country at all events, in new machinery and improvements in manufacture and agriculture.

6. The British sugar industry remains much where it was, waiting to know, as our Allies are waiting to know, what the British Government is going to do, and until that is settled, nothing will be done to develop the industry in British territory. Only one new factory of any size has been installed in the British West Indies for many years past, viz. that in St. Kitts. There has, so far as I know, been no fresh investment of capital in developing new land with a new factory of any size.

7. The capital expenditure required to make the Empire self-supporting in sugar will be some £40,000,000 to £50,000,000, a sum about equal to the excess cost of sugar to this country since war began. This cannot be tempted into the industry without adequate security.

8. There has been in the past no inducement in this country for young men to become proficient chemists and engineers for sugar-making. Increased facilities for education of this type have to be provided, and the inducement to make use of those facilities,

9. The agriculture and manufacture of sugar must be improved. Research work must be carried on, on a larger scale. I have already pointed out the magnificent results achieved in Germany by prudent and beneficent legislation in this direction. The same may be said of U.S.A. colonies and dependencies. I do not go so far as the writer of the book, *The High Price of Sugar and how to Reduce it,* in which a statement is made to the effect that but for the obstinacy of the planters in West Indies, they might be getting nine tons of sugar to the acre, as stated to be the average in Hawaii. I have reason to believe that the present average in those islands is nearer five tons per acre.

The question of the best economic yield per acre is a matter of great difficulty. I have no hesitation in saying that the means adopted to bring about the high yields obtained in Hawaii, the highest known yields of cane sugar in the world, could never have been achieved in the British West Indies under old conditions with profit to the producer, and I doubt whether it is desirable to create conditions which justify such extravagant cost of production. It is certain that if the Act establishing free import of sugar to the United States in 1916 had ever been brought into force, the Hawaiian yield would have been reduced, as they would no longer have been able to afford the intensive and expensive methods which have, under a protection of £8 per ton, secured such high yield.

On the other hand, the complete absence of those conditions and the lack of security for the industry in the British West Indies has resulted in low yields. No doubt under more favourable circumstances, both the quality of the cane and the efficiency of the extraction from the cane could be considerably improved. But where there are immense areas of suitable land available for the production of sugar cane, as in the British Empire, it is, as I have already said, a matter for careful and educated consideration at what yield point to stop. In Cuba the yield of sugar per acre is low, a little over two tons. Manure is only beginning to be applied to the older estates, though the factory work is of a high standard. In Java and Hawaii the yield per acre is high, over four tons per acre. Hawaii, Java and Peru raise sugar by irrigation; Java is fortunate in having 30,000,000 natives, and therefore plenty of cheap labour.

There is a vast field for research and improvement in the breeding of suitable canes, their proper treatment in the field, the investigation of the pests which attack them and the remedies, the treatment of the soil, drainage, irrigation, etc., which hitherto have not been able to be carried out under British conditions.

All these points, excellent and worthy of attention, demand security for the capital required for the necessary research, and which can only be obtained by beneficent Government legislation. Take India, for example. Sugar is produced there under conditions almost the most primitive in the world, the yield being about one ton of sugar per acre. A very moderate application of scientific methods to her sugar industry would soon make her an exporting instead of an importing country, but there is no inducement under Free Trade conditions.

The following statements which have been carefully

	Hawaiian Islands	Queensland	Java		Cuba		Mauritius	Demerara	Porto Rico	Louisiana
	1908	1915	1913	1906	1915–16	1914–15	1915	1909–15	1914–15	1914–15
	32 Mills	1 New Mill (Babinda)	133 Mills	• 96 Mills	9 Mills	Chaparra Mill	28 Mills	1 Mill	2 New Mills	1 Modern Mill
Sucrose % Canes	14·03	14·75	12·54	12·42	14·10	12·40	13·82	12·33	12·74	10·10
Fibre % Canes	12·37	9·02	12·40	11·48	11·04	11·00	12·55	12·20	12·06	11·46
Purity Clarified Juice	87·75	83·6	82·5	85·9	84·2	83·00	85·1	85·1	84·9	79·6
Extraction. Juice Sucrose % of Cane Sucrose	93·55*	93·3	90·7	90·6	†91·1	92·8	90·2	90·04	92·6	86·1
Sucrose obtained % Sucrose in Juice	92·20	76·9	85·7	88·8	‡88·3	85·5	86·6	87·90	91·9	88·0
Sucrose obtained % Sucrose in Canes	86·25	71·8	77·7	80·5	§80·44	79·4	78·11	79·1	85·1	75·8
Sucrose in Sugar % Canes	11·96	10·58	9·75	10·1	11·34	9·84	10·79	9·75	10·84	7·65
Tons Cane per ton Commercial Sugar	8·18	9·29	10·0	9·6	8·46	9·85	x8·94	9·91	8·90	12·73
Polanization of Sugar	97·8	98·3	96·5	—	95·9	96·9	x96·5	96·6	96·5	97·4

* *Av. Extraction.*
1913, 94·2.
Ewa 1913.
Extraction 97 %.

Av. for Cuba.
† 83·6.
‡ 89·0.
§ 74·4.

compiled from latest available data, give the manufacturing statistics of various sugar-producing countries, and it will be seen that those for British countries do not compare unfavourably with others, except Hawaii and Porto Rico. In Porto Rico land is very dear owing to competition from tobacco growing, and in Hawaii all land that can be irrigated has been taken.

The following records of progress in the chief sugar-producing countries illustrate the necessity and value of beneficent legislation in the development of the industry—

JAVA.—The yield expressed in tons per acre has steadily increased.

In	1840–44	yield	per acre	was	·809	tons sugar.
„	1865–70	„	„	„	1·72	„
„	1877	„	„	„	2·34	„
„	1888	„	„	„	3·25	„
„	1900–1	„	„	„	3·12	„
„	1910–11	„	„	„	4·30	„
„	1915	„	„	„	3·50	„

The progress latterly has been chiefly in agriculture, the yield per acre in tons of cane having increased in the seventeen years from 1893–4, 27·108, to 1910–11 41·945 tons per acre, an increase of nearly 50 per cent. In 1915 the tonnage of cane per acre was 38·25. The yield of sugar on weight of canes in 1910–11 was 10·26 per cent.; in 1915, 9·15 per cent. Mr. Prinsen Geerligs states the pre-war cost of production to have been about £7 10*s*. per ton f. o. b. Java.

CUBA.—The yield in tons of cane per acre over a period of years is not available, so that it is not possible to estimate the improvements in agriculture. The yield per acre in 1915–16 was stated to be 2·15 tons sugar per acre. The progress in sugar produced is given in recovery of sugar on weight of cane. This was—

In 1902–3 . . . 9·54 per cent.
„ 1910–11, 11·35 per cent. as against Java 10·26 per cent. in 1910.
„ 1915–16, 11·52 per cent. „ „ 9·15 „ „ in 1915.

Up to the present time no manure is used, except on the older estates, but it is stated that the immense extensions recently made in factory capacity will necessitate more intensive cultivation in future.

The developments during the war which have been and are being carried out, and from which British producers have been debarred by excess profits and income tax, put Cuba in a very strong position to improve her yield and extraction. The importance of this as affecting producers in the British Empire after the war can hardly be overstated.

The cost of production is stated to average 2 cents to 2¼ cents per lb.—£9 6*s*. 8*d*. to £10 10*s*. per ton f.o.b. Cuba.

HAWAII.—Production has been brought to the highest pitch in agriculture, irrigation and manufacture under the £8 per ton protection afforded by U.S.A. Tariff. Tons of cane per acre have averaged as high as forty-nine in one year on the best estates, and sugar contents as high as 17 per cent. I saw in 1899 a crop growing which averaged nearly 10 long tons of sugar per acre, but the cost of production included enormous expenditure on manure and irrigation which would not have been possible in years of normal pre-war prices without the high protection of the U.S.A. Tariff.

The cost per ton is stated by Geerligs to be about £12 16*s*. 8*d*. per ton delivered U.S.A., of which £3 was freight in American vessels. The most notable point is the marvellous effect of irrigation. The returns on non-irrigated land in 1910 were 3·06 tons per acre against 6·27 on irrigated land. The canes are extraordinarily sweet, and the milling extraction said to be the highest in the world, over 97 per cent. in one case.

But the average yield of sugar per acre throughout the islands was in 1910 only 4·69 tons per acre, and is said to be about the same to-day.

The following are the latest figures—

	Acres in Canes	Acres Canes cut	Tons (2240 lbs.) Sugar made	Yield Tons Sugar per acre
1915	239,800	113,200	576,786	5·10
1916	246,332	115,419	529,253	4·59

Plant cane requires about eighteen months to mature. The area of canes cut for 1916 crop was 46·8 per cent. of the area of canes in cultivation; the remainder, 130,913 acres, was turned over to be crushed during the following year.

The average yields of cane per acre have been—

	Yield of Sugar % Cane.
1912, 42 short tons (2000 lbs.)	12·45 per cent. max. 12·0 per cent. min.
1913, 39 ,,	
1914, 45 ,,	
1915, 46 ,,	
1914, 45 ,,	
1916, 42 ,,	

The yield of sugar from the canes for crops 1912–16 has varied between the limits 12·0 per cent. and 12·45 per cent. The length of crop season varied during the same period from maximum 1912, 200 days, to minimum 1913, 169 days.

The great development began with annexation to the United States in 1898, when the crop was 204,000 tons, from which it has advanced to 575,000 tons estimated for 1916–17.

The above records show the possibilities of the industry under favourable conditions. The developments in Cuba and Hawaii were directly due to the Tariff policy of the United States; the development in Java to an inexhaustible labour supply at phenomenally low rates. Wages in Java used to average about 5*d. per diem,* but are now considerably higher. Wages in Cuba are over one dollar per day.

I should, however, emphasize the difficulty in drawing conclusions from above figures without the fullest information as to all the conditions. Sugar production varies in the length of time occupied by the cane in maturing. In Natal it takes two years. In Java, Cuba, Hawaii and the British West Indies over one year. In the West Indian Islands there is only one crop season. In Peru sugar is produced throughout the year; in British Guiana about nine months out of the twelve. In Cuba very little manuring is done. In Hawaii the pre-war cost of manure was $30 per acre, or over £6. In Java sugar is a rotation crop, always plant cane. In Cuba they grow many crops from the same stool. There is no industry that I know of in which figures are more liable to mislead and results more difficult to compare. But it may be taken as a generally accepted fact that 80 per cent.

return from 100 of sucrose in the cane contained in 96° crystals bagged is first-class work all over the world, and that the average yield of sugar per acre is under rather than over two tons to the acre.

I have now indicated what in my opinion will be the conditions in the sugar industry at the end of the war, and it only remains for me to put before you the measures which, in my opinion, it will be necessary to take to secure that the British Empire will in time to come be self-supporting in the production and refining of sugar as she was fifty years ago. It is for the Government of this country to decide whether they will take those steps, and for the people to say whether they will support them in doing so. I would, however, most respectfully urge that to transfer our dependence for our sugar supply—a staple food, and a basic raw material for many of our home industries—from Germany and Austria to Cuba and Java, or any other foreign country, will be to jump from the frying-pan into the fire.

National security demands that we shall never again allow ourselves to be found in the position of dependence on foreign supplies which we occupied at the beginning of this war. This has already been recognized in regard to wheat and other cereal production, and must be recognized in sugar. In the latter case, security fixed for a long term, say, ten years, is more necessary than in the former case, because the production and refining of sugar involves a larger amount of fixed capital in the form of machinery than almost any other industry. The British Empire Producers Organization held a conference last year of sugar producers from all over the Empire, and drew up a series of recommendations which, in their opinion, and I heartily endorse it, would bring about the necessary development. They took as their model the Canadian Tariff, which, in my opinion, is the most scientific tariff, and has had the best results of any tariff in the world. Their recommendations were as follow—

(1) That Empire Sugar be granted preferential treatment to the extent of 50 per cent. on any tariff that may be in force; such preference at no time to be less than one halfpenny per pound for sugar exceeding 98 degrees polarization, graduated according to the existing British Tariff.

(2) That sugar of their own production exported to the United Kingdom by Allied Countries may by negotiation be granted a concession not exceeding 12½ per cent. of the General Tariff in force.

(3) That sugar of their own production exported to the United Kingdom by Neutral Countries shall pay not less than the General Tariff rate, but that Neutral Countries declining to enter into favourable reciprocal arrangements with the Empire or giving more favourable treatment to other countries, shall in addition be subjected to a surtax.

(4) That the import of sugar, directly or indirectly, from Enemy Countries be totally prohibited for a period of five years after the war, and that thereafter it shall pay the General Tariff in force plus 50 per cent. surtax; such surtax at no time to be less than one halfpenny per pound for sugar exceeding 98 degrees polarization graduated according to the existing British Tariff.

(5) That the exporting country shall in every case furnish approved certificates of origin.

(6) That a dumping clause on the lines of the Canadian Tariff be instituted.

(7) That if at any time His Majesty's Government be satisfied that the export of sugar to this country is being assisted by means of bounties, cartels, rebates of freight, or any other artificial means whatsoever, immediate steps shall be taken to countervail such assistance before applying the tariff.

(8) That all products the composition of which includes sugar shall

be subject to the same scale of duties based upon the sugar contents.

(9) That the difference between excise on home grown beet and the duty on Empire grown sugar shall be £2 6*s.* 8*d.* per ton basis 98 degrees until the crop of home grown beet reaches 50,000 tons of sugar per annum, thereafter the advantage to cease.

(10) That the British Government be asked to adopt above recommendations forthwith, and to make Agreements with the Dominions, Colonies and Dependencies to secure preferential treatment of Empire Sugar on the foregoing lines for a term of ten years certain or such longer period as may be agreed upon.[1]

If our recommendations are adopted, I believe that we shall see in the British Empire a development which within the time mentioned, ten years, will go far towards, if it does not actually attain, the production required by the Empire.

There can be no question as to the capacity of the Empire to produce all its own requirements of sugar if once the inducement is given. The West India Committee instituted an inquiry into possibilities of expansion all over the Empire, and published the results of that inquiry in a pamphlet entitled, *The British Sugar Industry: a Memorandum regarding the Possibilities of its Development*, published in 1915, from which I quote the following figures—

[1] Our recommendations are based upon an assumed General Tariff on sugar of 1*d.* per lb. = £9. 6*s.* 8*d.* per ton. Should it be necessary for revenue purposes to retain or increase present duty £14 per ton = 1½*d.* per lb., we do not contemplate the preference granted to Empire and Allied sugar being increased beyond the ½*d.* and ⅛*d.* per lb. respectively, provided for in our recommendations. But the 50 per cent. surtax on Enemy sugar should be maintained, whatever the duty may be, and at no time should the difference between Enemy and Empire sugar be less than 1*d.* per lb. on sugar exceeding 98 degrees graduated according to the existing British Tariff.

	Present Crops Tons	Possible Tons
Barbadoes	35,000	51,000
British Guiana	105,000	2,500,000
Jamaica	15,000	35,000
Trinidad	50,000	87,000
Windward and Leeward Islands	26,000	67,000
Mauritius	244,000	305,000
Fiji	97,000	164,000
British East Africa	—	500,000
Queensland	217,000	500,000
Natal	91,000	332,000
	880,000	4,541,000

Nigeria, North and South Uganda, German East Africa, are all capable of and suitable for growing sugar. Egypt is not mentioned, but has a crop of 70 to 100 thousand tons, which could be largely extended. It is consumed locally. British India has a crop of about 2,400,000 tons, which is capable of great expansion by the application of modern methods. Great Britain is capable of a large production of beet sugar so long as it is found economically advisable to grow it. If beets were extensively adopted as a rotation crop to wheat, on the additional 3,000,000 acres which the Government has announced will be brought under the plough before 1918, a crop of nearly 1,000,000 tons of sugar might eventually be attained. It is therefore clearly established that there is ample land within the Empire suited by nature to produce not only enough sugar for the requirements of the British Empire, but almost of the world. Other conditions will determine which are the most suitable localities for the industry.

The estimates of the West India Committee, particularly with regard to British Guiana, have been characterized as Utopian because the crop of that colony for many years past has not exceeded 120,000 tons. Again the reason is simple. There has been no security for capital and no inducement to increase. The land is there, and unless more suitable land can be found more favourably situated, the one serious drawback, lack of labour, can and will, I am confident, be overcome. The labour supply is barely sufficient for the existing crop, but it is hoped that the present negotiations with the Government of India may result in the establishment of free immigration from that country capable of much greater expansion than the system whereby immigrants are recruited in small numbers under indenture. The possibilities of this colony depend mainly upon an adequate labour supply. At the same time, if Nigeria has equally suitable land, the labour exists there already in superabundance; and the fact that no sugar industry has previously been established there, is no reason why, under more favourable conditions, it should not be established and on a large scale. The one paramount necessity is that security shall be given to the industry. The indirect benefit to this country by such a development as is foreshadowed is enormous. Take the industry of sugar machinery alone. Away back in the 'eighties, the exports of sugar machinery from this country were large. They had fallen, prior to the Brussels Sugar Convention, to less than half, but the impetus given to cane sugar production all over the world, due to the abolition of bounties and the U.S.A. Tariff policy, has brought them up to the old level. Assuming an increase of 1,500,000 tons of sugar in the British Empire, the equipment required would mean an unprecedented demand for machinery. Then there are the subsidiary and allied industries of electoral machinery, irrigation machinery, railways and transport, fertilizers, etc. The prospects are immense, and cannot be overlooked in the general consideration of the question.

Reconstruction of the industry on modern lines in India alone would create immediate and large demands for machinery. Further, the importance of controlling the supply of sugar, which is the raw material of so many vastly important secondary industries, such as biscuits, jam, confectionery, mineral waters, etc., can hardly be over-estimated. These industries have been built up on a foundation of sand. They relied on foreign sources for their raw material, and have shared the fate of the ordinary consumers in the United Kingdom since war began. Their supplies and their production have been curtailed. They have almost a greater interest than other classes in advocating the development of a supply which shall be entirely controlled by British interests.

The following figures taken from the *Annual Trade Statement*, 1914, show the total value of the exports of sugar-using trades—

(*a*) Sugar and Articles containing Sugar: Confectionery, Condensed Milk, Biscuits and Cake, Cocoa.

	1910	1913	1914
To Foreign Countries	£2,030,996	£2,474,780	£1,967,961
„ British Possessions	2,692,119	3,296,701	2,865,935
	4,723,115	5,771,481	4,833,896
(*b*) Value of Beer and Aerated Waters.			
Foreign Countries	914,898	1,084,776	807,423
British Possessions	1,113,198	1,331,753	1,189,261
Grand Total	£6,751,211	£8,188,010	£6,830,580

It has been stated that our recommendations will unduly enhance the cost of their raw material, but I think this is a complete misapprehension. That sugar must be dearer in this country after the war than before is a certainty, because the revenue requirements due to the war will necessitate a high duty for years to come; but any development of the production within the Empire will unquestionably reduce the world's price, and the quicker the development the sooner they can hope for a return to more normal cost of their raw material. Further, the manufacturers will, under our proposals, receive preferential treatment in the Empire and Allied Countries for their products. The British Empire Producers Organization, in a letter dated November 21, 1916, to Lord Balfour of Burleigh, estimated that the sugar requirements of the Empire can be produced within the Empire at a cost to sell retail in this country of 2½*d.* per lb. exclusive of duty.

There is one more point that I wish to make before I close, viz.: Why is it that the production in the Empire has shown so little advance compared with that in other countries? No one can assert that we have not equally suitable land. The quality of beets grown in England shows that we can produce as high a grade as in Germany. Land eminently suitable for cane production exists all over the tropical possessions of the Empire. Therefore it is not lack of land. In the manufacture of machinery we are second to none, as is shown by the fact that factories and machinery produced by British manufacturers exist in all the best sugar-producing countries of the world.

It may be claimed that our education is not equal to

Diagram Illustrating Comparative Progress of Sugar Production, Consumption, per Head, and Price. British Empire, U.S.A., and Germany.

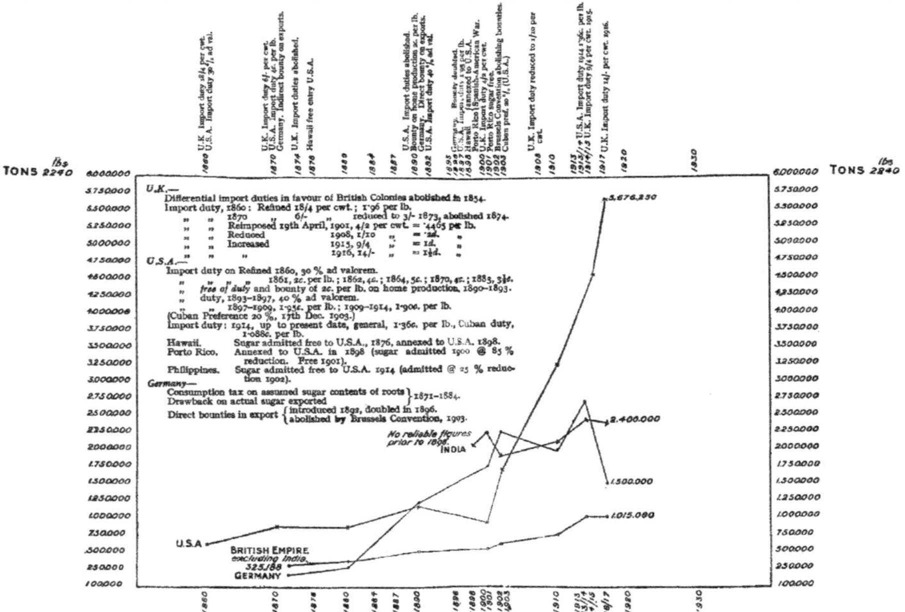

		lbs.	lbs.	lbs.	lbs.	lbs.	lbs.	lbs.	lbs.	lbs.
U.K.	Imports per head.	34	47	60	71	85	84	86	96	73
U.S.A.	Consumption per head.	29·6	32·7	39·5	54·56	66·6	72·8	81·6	84·29	79·34
Germany.	" " "		14·52	15·84	20·16	29·97	25·51	38·61	40·29	66·92
U.K.	Price refined, wholesale duty paid.	3·71d.	4·61d.	3·13d.	1·72d.	1·35d.	1·16d.	1·64d.		
U.S.A.	Price refined, wholesale duty paid.	9·78c.	13·51c.	9·80c.	6·27c.	5·32c.	4·455c.	4·972c.	4·685c.	6·862c.

U.S.A.—Domestic, Colonial and Cuban Cane and Beet, including Cuba, Hawaii, Porto Rico and Philippines
British Empire excluding India.
India. Germany. 1916/17, Estimated.

C. Sandbach Parker.
B.E.P.O.
2nd May, 1917.

that in other countries, but the most highly skilled and educated men can be obtained by paying for them from whatever country they are reared. Therefore we can only conclude that something is wrong with the conditions of our industry, and that this must be put right before any development can take place. Lack of security has been the sole cause of inferior development. This cannot be put right without a drastic change of policy by our Government.

One more point. Our whole policy in the past has been to maintain our position in the world as a manufacturing country by cheapness of production secured by cheap raw material and cheap labour. We allowed control of our raw material to pass into foreign hands, relying upon the power of our purse to secure it. We paid our labour low wages, and then were surprised when we had strikes, organization of labour against employers, and restriction of output in order to make the jobs go round. All the time our competitors were building up their manufacturing position at our expense, first by securing control of the raw materials, even those located in our own Empire, which we neglected, then by securing the profits of transport, and, lastly, by securing the profits of distribution even in our own country. They deserved to succeed, and we deserved to fail as we have done, and the war has brought it home to all of us that conditions must be changed and that we must never again neglect the control of raw materials which are the basic condition of all trade and prosperity. When once this country realizes, as all our Dominions have now realized, the vital importance of this primary question, we shall, I do not doubt, take the necessary steps to conserve and develop, not under the control or at the caprice of foreigners and for their benefit, but under our own control and for our own benefit, the natural resources which we have in such abundant supply.

The following passage from *The Times* of August 6, 1912, shows that members of His Majesty's Cabinet at that time were perhaps hardly fully aware of the importance of controlling our own supplies of raw material and necessaries of life—

"Mr. Hobhouse, Chancellor of the Duchy of Lancaster, speaking yesterday at Tenby, said that during the last few days the Government had denounced the Sugar Convention, which was brought into being by a Conservative Government. The Conservatives gave them to understand that the Convention would certainly keep down, if it did not diminish the price of sugar, but sugar had risen ever since. He was not curious to inquire whether the rise was the result of the Sugar Convention, or whether it was the result of some other factor. He was content to notice its lamentable effect. Before the Convention came into existence, thanks to bounties which not too wise foreign Governments paid upon the exports of sugar from their countries, England enjoyed cheaper sugar than the rest of the world, and they built up a great export trade in jams and confectionery, which they were sending to bounty paying countries. That trade had suffered a considerable set-back by the Sugar Convention, and it was high time that they took the opportunity of putting an end to the Convention which had been not merely a nuisance but a great blow at the prosperity of the people of this country. The Government had terminated this Convention and, he believed, to the great future advantage of merchants, trades and the consumer!"

Any thinking man will notice that the speaker was not curious, nay, anxious to find the reason for the rise in prices since the Convention has been in force, and that no figures were given to prove the considerable set-back in the export trade either. Apart from that, are not Mr. Hobhouse's prophecies entirely shattered to-day? Was the German Government unwise in giving drawbacks and the British Government wise in accepting cheap (bribed) sugar? The prophecy that the system of bounties would place us entirely at the mercy of foreigners, which was expounded by Baron de Worms (since Lord Pirbright) when he was Secretary for the Colonies, I believe about 1895, for which statement he stated in the House upon whom he relied, proved to be by far more true than statements of the Chancellor of the Duchy of Lancaster in 1912.

The recommendations of the British Empire Producers Organization with regard to sugar have been severely criticized in various quarters, including those of existing sugar producers of the Empire, but they represent the unanimous conclusions of the Sugar Conference, and I recommend them to your consideration. In doing so, I would allude to one criticism, by one of the sugar-buyers for the Co-operative Wholesale Society of Manchester, which has been widely circulated amongst the large membership of that society. Broadly he asserted that under our scheme the consumers of this country would pay £18,500,000 per annum, in addition to world's price to finance the policy of the sugar section of the British Empire Producers Organization.

The statement is grotesque, but requires some notice. Our scheme provides for a rapid increase of production in the Empire by means of a preference of ½*d.* per lb. for ten years. It is true that any duty imposed must be an addition to world's price, which will fall upon the consumer. It is certain that the import duty on sugar in this country must for revenue purposes remain much higher after than before the war for some years. But it is a fact that preferential treatment increases production in the countries receiving the preference, and that the consumers in the country giving the preference reap the benefit of this in the reduced world's price of sugar due to increased supplies. This is well illustrated in the case of the United States, where during twenty years, 1890 to 1910, there was a drop of twenty per cent. in the wholesale price of standard granulated concurrently with developments in production in countries preferentially treated.

	Cents per lb.		
In 1913 the average wholesale price for Standard Granulated was, in U.S.A. . . .	4·278	as against U.K. about	2·0 retail
Less duty	1·90	” ”	·2 ”
Value exclusive of duty 2·378	1·19 per lb.		1·8

The point that our critic seems to make is that present producers of sugar within the Empire will get an increased price for their sugar due to the preference. This is true only up to the point when Empire production is sufficiently increased to create healthy competition within its own borders, when the bulk of the benefit will be transferred from the producers to the consumers, as has been the case in America.

If the Empire were capable of at once supplying the whole requirements of this country, say 2,000,000 tons, a very small preference would be necessary to give security to the industry in the Empire, but this is not the case. A new industry has to be created, and we start where Germany started in 1871, and the United States in 1898. The British Empire Producers Organization estimate that only 400,000 tons will in the first year be available for this market on which the preference advocated would mean a loss of import duty to the Exchequer of not more than £1,866,000 or about $\frac{1}{10}d.$ per lb. on our consumption of 2,000,000 tons. If the effect of this advantage is to secure to the Empire in ten years its whole supply produced and controlled solely by British interests, it would seem a small price to pay, and it is certain that long before the full supply is reached, the benefit of the introduction of larger quantities at the lower rate of duty will be reflected in lower prices to consumers. This is proved by the experience of the Cuban and British West Indian Producers in the United States, and Canadian markets respectively, who during their heavy crop season get almost no benefit in the price from the preference, the whole of which goes to the importer and presumably reaches the consumer. It is perhaps relevant at this point to draw attention to the fact previously stated that the major part of the increased price to consumers in this country at the present time is due not to the increased duty, but to the increased world's price which would not have taken place had the Empire produced and controlled its own supplies. As already stated the price in Australia remains $3\frac{1}{2}d.$ and in New Zealand about $2\frac{1}{4}d.$ The same critic also made the following statement—

"As a matter of fact, the islands (B.W.I.) have been developing fruit and bananas on profitable lines and are to some extent letting sugar go. One factory in Cuba makes more sugar than the total of any single B.W.I. Island. British Guiana on the mainland makes less than twice as much as this single Cuban factory. Four such factories would make more sugar than the whole British West Indian possessions put together."

It is hard to think of any statement which would more completely condemn our sugar policy in the past, or more amply justify the British Empire Producers Organization's proposals for the future. I have seen that factory, I know its performances, and I say with confidence that given adequate security, the British Empire could and would possess many similar factories. The British West Indies and British Guiana were producing sugar long before that factory was thought of, and had they received similar treatment would have no doubt similar factories to-day. We want to see them, and we believe that the consumers in this country, Co-operative Societies in common with all others, would benefit by creating the conditions under which similar factories could be erected. So far the conditions do not exist, and it is absolutely certain, in my opinion, that unless some such policy as the British Empire Producers Organization recommend is adopted, we shall not only never see such factories in the British Empire, except in those countries which consume their own production, but we shall see the small and struggling industry which we retain to-day irrevocably crushed out.

We must in our own interests as consumers encourage production in the British Empire to the utmost. We have already encouraged it in Cuba : it is for our people to decide whether they will remain dependent on Cuba for their supplies in the future or encourage production in Great Britain and the Empire, as every other first-class country does, and as it has been laid down in the Paris Conference resolutions that the Allies should do. One thing must not be forgotten, viz. that while we in England thought the Germans were most unwisely making us a present of their sugar, they were not only making huge profits mainly at our expense, but had established a rural industry and were building up a race of splendid fighting men, also at our expense, to resist whom we have had to expend vast quantities of our best blood and treasure.

Messrs. Willett & Gray, eminent sugar statisticians in the United States, have lately made the statement that—

"Recently acquired knowledge and the experience of our countrymen have taught that the vast sum of money we are now sending to foreign nations in payment for sugar, might well remain at home, rural communities not only receiving the revenue they would expend in American markets, but also gaining indirect benefits through the increased productivity of soil following beet culture." The same applies with much greater force to the British Empire, particularly to the United Kingdom.

"Once bit, twice shy," is the motto I commend to the people of England. If they intend to prove themselves good citizens of the Empire, as our great Dominions have done, they must be prepared to treat Britons better than foreigners in future, and they will benefit by so doing.

Since this paper was delivered His Majesty's Government have announced, as the result of the deliberations of the War Cabinet, their adhesion to the principles of Imperial Preference, and to the policy of preferential treatment of our Allies embodied in the Resolutions advocated at the Paris Economic Conference.

The momentous importance of this declaration to the world at large and the British Empire in particular cannot be overstated. The Empire will breathe a sigh of intense relief, and our brave Allies will, now that they know the agreed policy of the British Government, be enabled to proceed with confidence to carry into effect the Paris Resolutions. But immediate executive action on the part of the Governments concerned is needed to enable us to face the trade position on the declaration of peace.

A great deal can and should be done now, for which no fresh legislation is required, and it will be the duty of all trade organizations such as the one over which I have the honour to preside, to press upon the Government the necessity of immediate action.

THE COTTON RESOURCES OF THE BRITISH EMPIRE

By JOHN A. TODD, B.L.
Professor of Economics, University College, Nottingham.
(Read on 16th February, 1917)

THE scope of this paper is mainly limited to the question of the supply of the raw material of the cotton industry, because it is no exaggeration to say that no man alive knows the whole cotton trade of the Empire, let alone of the world, from raw material to finished product. The cotton trade is so enormous and so highly specialised that he would be a bold man who professed to know even the whole of the Lancashire industry alone. All that the writer can claim is an outside acquaintance with the various branches of the industry in the United Kingdom, and ten years' intermittent experience (also from outside) of cotton-growing in Egypt, America and India; for the rest—and the cotton-growing areas of the world are scattered over practically the whole of the tropical and sub-tropical regions of the globe—his information is necessarily second-hand.

Such a limitation of the scope of the inquiry would, however, be justified to a degree probably greater than in other industries, because in the cotton trade the value of the raw material is a large percentage of the cost of the finished article, and the problems of its supply play an important part in the general position of the industry. According to the *Census of Production* 1907, the total output of the trade in Great Britain was about 1800 million pounds' weight of yarn, the value of which averaged about a shilling per pound. To produce this amount of yarn required about 2200 million lbs. of raw cotton, which at that time averaged about 7*d*. per lb. in value. Thus the average cost of the raw material represented about 8½*d*. per lb. of yarn produced, or over 70 per cent. of the value of the yarn, while the yarn again forms probably a similar proportion of the value of the finished cloth.

Again, the writer must necessarily treat the subject from the point of view of the economist, not that of the botanist; [1] but that also is justified by the fact that the present-day problems of the Empire's cotton supply are at least as much economic as botanical.

The importance of the cotton trade as one of our staple industries has been so frequently enlarged upon that it need only be briefly referred to here. Cotton is our largest single import of any commodity, and averages fully 10 per cent. of the value of our total imports. Cotton goods again are by far our largest export, representing from one-fifth to one-fourth in value of the whole. Taking the figures of our export trade in 1907 and comparing them with the estimated total output according to the *Census of Production*, it appears that the export represented about nine-tenths

[1] For this aspect of the subject see "Cotton" by Dr. W. Lawrence Balls in the preceding volume of the Imperial Studies Series, *The Exploitation of Plants*, edited by Prof. F. W. Oliver, F.R.S.—Ed.

in bulk of the output of the trade, but only about eight-tenths of the output in value, at a rough calculation, so that our home consumption is evidently enormous, say about £30,000,000 in value. Tables A and B in the Appendix give a few recent figures bearing out these estimates.

With regard to the position of British industry in the world's cotton trade, Great Britain owned in 1912–13 (the last year for which complete figures are available) 39 per cent. of the world's spindleage. To this India added 4·2 per cent., and Canada 0·6 per cent., making the Empire's total share about 44 per cent. of the whole. Our consumption of raw cotton in weight was, however, quite out of proportion to these figures, the United Kingdom taking only 18·6 per cent. of the world's total, while India took 9·5 per cent. and Canada 0·5 per cent. The explanation is, of course, that the class of cotton used in Lancashire is, on the whole, the finest in the world, while India is at the other end of the scale, and the weight of cotton consumed is in inverse proportion to its value (see Table C in the Appendix). This, again, affects the relative value of the manufactured products of the Empire's cotton industry and its various parts. The value of the Lancashire product is probably well over 50 per cent. of the world's total, while the value of India's product is probably even lower in the world's total than is represented by her proportion of its spindleage. The question of the Empire's share in the production of the raw material throughout the world will be dealt with later on.

The supply of its raw material is probably the greatest problem which the cotton trade has to face to-day. The trade has just passed through a crisis in this respect, which a few years ago would have been regarded as sufficient to ruin the industry. The central fact of the crisis may be seen in the price of the staple, which has touched records unequalled since the days of the cotton famine of the 'sixties. "Shilling cotton" has been for over forty years a thing which no one expected ever to see again, and though during the last year or two some people have talked of its possibility, they found few to listen. But the shilling limit has been passed with a considerable margin, for on November 20, 1916, the price of "Middling" American cotton, which is the basis price of cotton throughout the world, touched 12·59*d.* per lb. in Liverpool.[1] Such a height has not been reached since 1869, and since that date the opposite extreme has been touched on two occasions, in 1894 and 1898, when the price was below 3*d.* Even in 1904 Sully's famous corner only forced the price up to 8·96*d.*, and that did not last long. As will be seen from Table F the average price for some years before the war had been round about 7*d.*

It is, however, a striking fact that the price of cotton has not in reality been at all extraordinarily high when compared with pre-war levels and the rise in the prices of other commodities. Even the high price of November 1916 compared not unfavourably with the general level of prices as a whole as shown by the *Economist* Index Numbers, and the rise in the case of cotton was distinctly less than that of the prices of textile raw materials as a whole. What has given such prominence to the high price of cotton is, that it has risen from a

[1] See P.S. on p. 108.

very low point, due to the fact that the first effect of the war was to cause a heavy fall in the price, so that the recent high price seems abnormal when compared with the opposite extreme of two years ago. Both the fall and the rise were due to war conditions, and these must now be briefly described in order to make the critical position of the industry plain. The essential fact of the situation which makes it so threatening is that the war has merely accentuated certain unsatisfactory features of the pre-war position of the industry, and that the cumulative effect is likely to be more or less permanent.

The basic fact of the whole situation is that the world's consumption of cotton had been increasing for at least ten or fifteen years before the war, at a rate which was only limited by the actual supply available. The increased demand for cotton goods of all kinds has been due to the steady increase of the world's population and the still more marked increase in the prosperity of that population, especially in those countries such as India, where large quantities of raw materials for the world's industries are produced. It so happens that these countries are also just those in which clothing is more or less of a luxury, and is most universally of cotton. Thus the increase of prosperity with the consequent improvement of the standard of living has shown itself in the greater demand for cotton fabrics of all kinds, for it must be remembered that even in temperate climates cotton forms a large part of the world's clothing, and that on the whole probably nine-tenths of the fabrics employed are of cotton. Again, the demand for cotton has been largely due to the successful substitution of cotton as a cheaper raw material for the manufacture of all kinds of fabrics which were formerly made of wool, linen or silk. At the same time new uses for cotton have been continually cropping up, from the webbing of motor-car tyres to typewriter ribbons and aeroplane wings; and the possible developments in this direction seem to be unlimited. But probably the main factor in the increased demand for cotton is the wonderful improvement in recent years of the processes applied to cotton, as the result of which cotton goods are now produced of a quality and finish which have entirely altered the comparative place of cotton in the scale of values. Cotton is no longer a term of contempt; it is a really satisfactory substitute for more expensive raw materials; and the fabrics which can now be produced from pure cotton are such as to satisfy the most fastidious taste. The result is, that it is now possible to obtain at the most moderate prices fabrics equal in appearance to those which a generation ago were only available to the richest buyers because of their high price. This has brought such fine fabrics within the reach of practically every class, and as the demand thus touched is now a middle class demand it is enormous, and, indeed, almost insatiable.

At the same time, as will be seen from Table D, the world's supplies of cotton have been increasing rapidly; but the increase has not kept pace with the potential increase of the demand, with the result that prices have been, on the whole, rising steadily, and, as has been said, the consumption has only been limited by the supply. For ten years before the war the season's consumption was actually in excess of the year's crops in five years out of the ten (see Table E). It is, unfortunately, impossible to continue these figures for the period since the war, as consumption statistics are, of course, no longer available from the German and Austrian sections of the industry, but the figures of the American crop and consumption are available, and these, which will also be found in Table E, show that the state of affairs since the war has been substantially the same. As the American crop dominates the whole world's supply, this means that the whole position since the war has been at least as bad as it was before, so that the question of an increased supply of raw cotton for the world has become still more pressing than it was before the war.

This dominance of the American crop over the world's supply, of which it still forms about 60 per cent., is the next feature of the situation which requires emphasis. The amount of the American crop every year practically rules the world's prices of cotton of all kinds, and, unfortunately, things have not been going well with the American crop. The main fact of the situation is that owing largely to the rapid increase of the labour cost in the American Cotton Belt, the cost of production of cotton, which requires highly intensive cultivation and an immense amount of hand labour, especially at particular seasons of the year, has been rising very rapidly. This is specially true in certain large sections of the Belt, such as Texas, whose crop of nearly 5,000,000 bales per annum forms so large a proportion of the American crop, and, indeed, of the world's total crop, that it cannot be dispensed with. In the course of a comprehensive tour through the Belt in 1913 the writer went into this question in some detail and satisfied himself that taking into account the full economic cost of production, including interest on, and depreciation of capital, the full labour cost at market values, and the cost of supervision, the cost of production in large areas in Texas is now so high that unless the price of American Middling in New Orleans is round about twelve cents per lb., the field price received by the planter is not sufficient to remunerate him adequately.

This has produced a remarkable result. A considerable section of the industry being pretty close to the margin of profitable cultivation, the effect of the price upon the acreage sown each year has become very noticeable, with the result that a sort of vicious circle has been established in the relation between area, crop and prices as follows:—(see Table F). A good year means a big crop, which means low prices; this causes a reduction in the acreage of the following year, and with ordinary weather conditions a reduced crop. This sends the price up again, and next year the acreage goes up once more. This see-saw motion is, however, frequently disturbed by the vagaries of the American climate, which may be seen in the fluctuations of the average yield per acre for each season in Table F. Thus a big acreage may not mean a big crop, if the season's weather happens to be bad; or a small acreage may give a fairly good crop, if the weather is specially favourable. But the general state of uncertainty, and the frequent and violent fluctuations of price thus produced, are very bad not only for the consumers but also for the producers of the staple. These peculiar conditions may be further illustrated by what has happened since the war as follows:—

The year 1914 was a record year as regards the American crop. The acreage was almost the largest on record, for, though the crop of the previous year had been fairly large, it had been well taken up at a good average price, which went far to compensate for a yield slightly below the average (see Table G). Thus 1914 began with good prospects as regards price, and the area put under cotton was quite satisfactory. The season's weather proved unusually favourable, and early in the summer it became evident that a record crop was assured. At the same time demand showed signs of slacking off after the boom years of 1912–13, and with the prospect of a big crop prices were naturally sagging a little. They were still good, however, when the war came suddenly upon the world, producing an alarming change in the prospects of the cotton trade, which seemed likely to be harder hit than almost any other industry. The result was a serious slump in prices which was not fully represented even by the steady fall of the nominal prices fixed by the committees of the various exchanges. The official minimum price in Liverpool fell to 4·25*d.* in December, but cotton was actually selling in the South as low as six cents. Such a price, of course, meant practically ruin to many of the planters; but the South, as a whole, rose to the occasion and displayed a holding power which no one had foreseen, with the result that prices soon rallied and maintained a fair level during the spring of 1915. There was, of course, a very strong movement not only throughout America but in all the other principal cotton-fields of the world, to secure a reduction of acreage in 1915, and the result will be seen in the reduced figures shown in Table G. During the latter half of 1915 demand began to recover to an extent much greater than any one could have hoped for, and prices rose accordingly during the winter. Unfortunately, however, the rise came too late, or its permanence was not sufficiently assured [1] to have the desired effect upon the average for 1916, which showed only a moderate increase upon the figures of 1915, and was still far below the 1913 records in almost every country. Then early in the summer of 1916 the market began to realise that the situation was uncomfortably near shortage. The demand for cotton goods of all kinds throughout the world was advancing at a most disconcerting pace; and it became evident that not only had 1915 eaten very largely into the surplus of 1914, but the whole of the remainder would hardly be enough to meet the demand of 1916–17 season, unless the 1916 crop were, at least, up to average. Hard upon the back of this came disquieting rumours about the weather in America, and very soon it became evident that nothing short of a disaster was in prospect. Prices at once began to rise rapidly and soon got beyond all bounds. The world was caught hopelessly short of cotton, and there was simply no limit to the price except the ability of the consumers to pay. That apparently is almost unlimited now, and prices have remained high ever since.

It will be seen that there is nothing in all this to surprise any one who had the facts before him, and was able to apply ordinary economic principles and common sense to the situation. As for the future, it is, of course, largely speculative as regards demand, but assuming

[1] There was a marked break below 12 cents in February, 1916.

that that remains anything like as active as it has been during 1916, there is little reason to hope for an early return to anything like pre-war prices. The acreage in 1917 ought to be record-breaking, though it was most unfortunate from this point of view that there was again a very marked break in prices during the present month, which may seriously check the tendency to go back to cotton. But judging from the way in which consumption, especially in America, has stood the test of record prices, it seems dangerous to assume that anything less than an absolute bumper crop will meet the requirements of the coming season, especially if the war comes to an end; and everything depends on whether such a crop will be secured. There are many possibilities against it—the temptingly high prices of alternative crops such as cereals, especially if America comes into the war, the lack of fertilisers, the increasing scarcity of labour, and the spread of the boll-weevil. We shall, therefore, need the best of luck in the weather for 1917, if the world is to be saved a repetition of the conditions of 1916; and we may quite easily see shilling cotton again.[1] Indeed, it may be safely prophesied that for some years to come the normal level of the price of cotton in this country, based upon the twelve cents level of cost of production, which has certainly not fallen since the war, and the high freights which are sure to continue for some time after the war, will quite likely be round about 9*d*., with an occasional run up to a shilling again every time that the American climate chooses to disappoint us, and the crop is short. Such a prospect is a very serious matter for the cotton industry of the world. It means

[1] See P.S. p. 108.

in effect that the world is more or less permanently short of cotton, that the predominance of the American crop and the conditions of its production are liable to cause very undesirable fluctuations in the supply, and therefore in prices, while owing to the inability of the supply to increase as fast as the demand could take it up, and the rising cost of production in America, the general level of prices must, on the whole, remain high, for only a very high price will suffice to maintain the necessary increase of the American crop. What the world wants is, therefore, a steady increase of about a million bales per annum, and this increase must be *cumulative*.

It is, then, a matter of urgency not only to the world at large as consumers of cotton goods, but to the world's cotton industry, in which the British Empire has so large a share, that the supply of cotton should be increased. It is also very desirable that the new supply should, as far as possible, come from some other part of the world than the North American continent, so that the weather conditions, upon which the annual yield must largely depend, should not be the same as those in the American Cotton Belt, thus spreading the risks of the crop and minimising the chances of a complete world failure in any one season. At the same time it would be well if the increased supply could be found in some country where, owing to a larger supply of the necessary labour, the labour cost of the crop need not be so high as it now is in most parts of the American Belt. Cotton has always been regarded as a cheap labour crop, and labour in America is no longer cheap. Finally, if the increase could be found somewhere within the British Empire, it would be the best thing possible for us. The next thing, therefore, is to consider the present sources of the world's supply of cotton and what prospect there is of the desired increase.

Table H in the Appendix gives in condensed form a summary of the world's chief sources of supply of cotton, classified in five grades according to quality, and showing the approximate quantity of each grade. At the same time the proportion of each grade grown in the British Empire is indicated. It must be explained, however, that this division into grades is the writer's own, and is not a recognised trade classification, but something of the kind is extremely convenient. It may also be explained that the value of cotton depends mainly upon the length of its staple, for upon that depends the "count" or fineness of the thread or yarn into which the cotton can be spun. Thus the longer the staple the finer the count of yarn and the higher the price. The best Sea Island cotton, for example, which has a staple of over two inches in length, will spin into yarn of a count as high as 300's, which means that 300 hanks of the yarn, each of 840 yards in length, will weigh only one pound avoirdupois, or, in more graphic terms, that about 150 miles of this yarn would only weigh one pound. Such cotton is naturally of enormous value, and before the war the best of it was worth about 3*s*. 6*d*. per lb. The total amount of these special crop "lots" is extremely small, and it is only used for very fine sewing cottons, and for the finest cotton goods. At the other extreme, the bulk of the Indian crop is of only three to five eighths of an inch in staple, and will hardly spin above 10's, while its value at the same date would be about as many pence as the other was shillings.

The main point of interest in this table is to see how the Empire stands with regard to the production of the various grades. The position may be summed up as follows. In the three highest grades we stand very well, though the maximum possibilities of development in these grades have by no means been reached. Again, at the other extreme we control the great bulk of the supply of the lowest grade; but in the medium grade which forms the bulk of the world's supply, the American crop is practically 90 per cent. of the whole, and our share is insignificant. From the point of view, therefore, of the Lancashire cotton industry, what is wanted is, first, a steady increase of the Empire's supply of fine cotton, such as Sea Island from the West Indies and Egyptian cotton from Egypt and the Sudan; and, second, a really substantial supply of good "bread and butter" cotton of about American quality from some part of the British Empire. If, for example, a million bales of good inch staple cotton could be grown in India, either in addition to or in place of some of her present very low grade crop, or in some other part of the Empire, such as Africa, it would just about serve the purpose. What the world wants is more cotton. What Lancashire wants is more *good* cotton, and if it can be found in the British Empire so much the better.

What, then, are the prospects? The answer involves a rapid survey of the conditions of production in all the main cotton-growing countries throughout the world, and it will probably be most easily followed if the survey, instead of being in geographical order, is taken from the point of view of the quality of the crop as indicated in Table H. Thus beginning at the top, the very best

Sea Island cotton is only grown on certain islands off the coast of South Carolina, near the port of Charleston, and in the West Indies. Next to this come certain lower grades of Sea Island, known as Floridas and Georgias, because they are grown in the landward districts of these states not far from the sea, and from original Sea Island seed, which, however, rapidly deteriorates and has to be frequently renewed. The future of these districts is unfortunately very seriously threatened by the approach of boll-weevil, which is now well into Georgia and will almost certainly wipe out the Sea Island variety in these areas; for these staple cottons are almost invariably slower in their growth and later of coming to maturity, so that they are peculiarly liable to the attacks of boll-weevil, which does its worst damage to the late maturing bolls. It is almost certain, therefore, that the fate of Sea Island cotton in these districts will be the same as was that of the old long-staple varieties which used to be grown in the Mississippi Delta, and which simply disappeared in front of the boll-weevil invasion. It is doubtful whether even the Islands themselves will escape the pest. If they do not, then the world will be thrown back for its supplies of the very finest cotton upon the West Indies alone, and it is doubtful whether the area available there will be equal to producing a sufficient increase of supply. Certainly the matter is of the greatest importance to them as well as to Lancashire.

As for the second-grade Sea Islands, we ought to have been ready to take their place at once by an extension of the best varieties of the Egyptian crop, such as Sakel and Jannovitch, which have for some time been keen rivals of Florida and Georgia. Unfortunately, however, the position of the crop in Egypt has during the last few years become extremely unsatisfactory, all the more so by contrast with the very bright prospects which were held out a short time ago. The extension of the area under cotton, and the improvement of its quality by the introduction of new varieties, had been making steady progress for many years, and had not by any means reached their limits. Projects were under consideration for the drainage and reclamation of large areas in the north of the Delta, which the extension of the irrigation facilities had at last rendered possible; and other irrigation schemes were on foot in the Sudan, which would not only have added to the possible area in Egypt, but would have brought into cultivation large areas in the Sudan itself, such as the Gezira, where the prospects for cotton-growing were excellent. Unfortunately, however, the war stopped all these projects, and so far from the crop continuing its advance it has since 1913 gone seriously backward (see Table H). This has also been partly due to the invasion of the country by a new pest known as the pink boll-worm or seed-worm (not to be confused with the American boll-weevil), which has been allowed to get a firm grip of the crop and will now be exceedingly difficult to dislodge. This is all the more to be regretted because the demand for Egyptian cotton not only of the very best varieties, but also of the ordinary Brown Egyptian and Upper Egyptian varieties, had never shown the least sign of slackening. There was room in the Lancashire market for all the additional cotton Egypt could grow, quite apart from the possible failure of the Florida and

Georgia crops; and the recent failure of the Egyptian crop to maintain its pre-war records, let alone its former rate of increase, may prove a very serious matter for the trade, as well as a calamitous loss to Egypt herself. In the Sudan the loss is rather of a negative kind, in that the war has still further delayed the necessarily slow development of the country; but the loss is hardly real in one sense, because, next to irrigation facilities, the greatest need of the Sudan for the development of a large cotton area is population, and that can only come with time.

Now that the first effects of the war on the finances of Egypt have disappeared, it is absolutely necessary that these large projects should be proceeded with at once. Every year's delay in their completion will mean the loss of many millions per annum to Egypt after the war. Time was when such matters in Egypt were rather a matter of politics than of engineering or finance, but now that Egypt is part of the British Empire, and the absurd difficulties of the Capitulations are to be removed, there is no excuse for the neglect of any measures which will remedy the present unsatisfactory state of affairs in Egypt. What is perhaps most to be regretted in Egypt is that until a few years ago there was going on under the Department of Agriculture the most promising work on the scientific improvement of cotton, by breeding and selection, but this has now been practically abandoned just at the point when it was coming within sight of fruition. The result is, that the old characteristic deterioration of Egyptian cotton is once more going on apace, and there is no indication of anything being done to counteract it. Egypt, which used to be the model to all cotton-growing countries, is now in danger of becoming, like so many others, a land of great possibilities but of relatively poor performance.

Moving southwards into Uganda one is just passing over the line from Grade III into Grade IV. The Uganda cotton is of the type known as American Long Staple, being descended from some of the varieties which made the Mississippi Valley a centre of improved cotton-growing up till a few years ago. The success of these staple American varieties in many other parts of the world has been very striking. In some parts of the Sudan they promise to be even more successful than the Egyptian types. In Sind in India they have also been successful where the conditions did not suit Egyptian, and in practically all of the new fields where the British Cotton Growing Association has been doing such excellent work in East Africa, they have been the most hopeful of all the foreign varieties tried. Thus in Uganda there is now a well-established cotton-growing industry, but unfortunately it is still on a small scale,[1] and there is no immediate hope of a really large increase. The main difficulty is the unwillingness of the native to abandon his natural habits of indolence or irregular work in favour of the sustained industry of agriculture; while the long transport to the sea is also a heavy handicap, and in many districts the cotton has to be brought by carriers to the rail over great distances. The lack of communications, especially railways, is also the chief obstacle to development in Nyasaland, where the prospects otherwise are most excellent. It will take a good deal of time and much money to open up these districts;

[1] The record crop was 42,000 bales in 1914.

but there is no doubt that both in these areas and in others farther south, such as Rhodesia, Natal, the Transvaal and the Cape Colony, there are great possibilities for the development of cotton-growing. Unfortunately the prospects of an immediate and substantial increase are completely lacking. All the work of the European Associations, German, French, Italian and Portuguese as well as British in every part of Africa, had not up till 1913 succeeded in producing a total crop of 100,000 bales in a year, and the fact must be faced that while the capabilities of these areas have now been proved beyond doubt, there is still much to be done before these pioneer successes can be translated into terms of, say, a million bales of cotton.

Coming now definitely to Grade IV, the great bulk of the world's crops, which in effect means the American crop, it has been seen that the unsatisfactory position of that crop with regard to future increase is the cause of the whole trouble. It would be a mistake, however, to imagine that the American crop has reached its limits either in respect of improved quality or increased quantity. It is rather a question of cost, especially labour cost and price. The crop could be doubled if the labour were available, and the price obtainable were sufficient to pay for it. The quality of the crop as a whole could also be much improved, if the best varieties obtainable were generally used wherever they are suitable. In many districts the Government Departments of Agriculture (State and Federal) or enterprising private growers have shown what can be done towards improving the staple and yield of the ordinary American crop, while a great deal has been done towards replacing the old long staple varieties by others, which while sufficiently early maturing to escape the worst ravages of the boll-weevil are still of long enough staple to command a good premium and very heavy yielders. There is great scope for development in this direction in every part of the Belt; and it is probable that economic conditions will gradually force upon the best American planters the necessity of doing something to improve the yield and quality of their crop, for only in that way can they meet the increasing cost of its production. There are many things to be done, however, before the character of the American crop, as a whole, can be materially improved, and judging from the apparent unwillingness of the trade in America to tackle the still more immediate question of the improvement of baling methods, which have been crying out for reform for many years, it is hardly probable that any great change will be made in the general conditions within a short time, or any time which would be of much avail from our present point of view. The question for us, therefore, in the meantime is whether there is any other part of the world, and particularly of the Empire, where an increased crop of cotton of this grade could be hoped for within a reasonable time.

Outside of the United States and the British Empire there are only three areas in the world of any importance from this point of view, namely, South America (with which Mexico may be bracketed), China and Russia. The first of these is almost out of the question, because Brazil, which has the largest crop—about 400,000 bales—is now consuming a very considerable portion of it herself, and is likely to do so still more in future as her own mills develop. There is no doubt that Brazil could produce a very much larger crop than she does, and much of it of fairly good quality, but most of the necessary conditions for such an increase are absent, *e. g.* labour, capital and security. The conditions in Mexico are similar, but much worse. Peru, again, grows some very fine cotton, some of it quite of Grade III quality, but the possibilities of extension are limited by the need of capital for irrigation works, etc. The Argentine could also grow a large cotton crop as far as climate and soil are concerned, but the scarcity of labour seems to be an absolute bar to its extension.

The Russian crop grown in Russian Turkestan and Transcaucasia is one of rapidly increasing importance, and its existence has probably saved the Russian industry from extinction during the war. As a matter of fact it is the only crop which is known to have increased since the war, and it now amounts to probably 1,500,000 bales.[1] A considerable part of it is of about American quality, but the native varieties are more like the inferior grades of Indian cotton.

China is the great mystery of the world's cotton supplies. There have been much exaggerated reports of the amount of the crop; but the most recent opinions —there are no statistics of any value whatever—seem to indicate a crop of about two million bales, of which the bulk is used in domestic consumption, for wadding in garments, etc. A certain amount is used in the local mills, and less than a quarter of a million bales are

[1] The latest information gives much less satisfactory news of the Russian crop which has been seriously reduced owing to the local need of cereals, high labour cost and other internal troubles.

exported to Japan; but the possibilities of extension of the crop seem to be very great. American varieties have been tried with apparent success in some districts, but the bulk of the crop is probably more like the Indian varieties and of comparatively short staple.

Turning now to the British Empire, there is only one country from which we might hope for a substantial increase in the immediate future, namely, India. The Indian crop is the second largest in the world, though last in point of quality; but its size is quite out of proportion to the enormous area it covers—25,000,000 acres, or fully two-thirds of the total area under cotton in the United States (see Table G). The average yield is probably the lowest in the world, and it is in this respect that India offers the most obvious room for improvement. A very slight annual increase in the average yield per acre would give the desired increase of a million bales; but there are unfortunately great difficulties in the way. The methods of the cultivators are primitive, and they are very slow to alter them unless an immediate profit is clearly shown by the change. The same applies to any question of the extension of the area under cotton. There is no doubt much room for cotton-growing in new districts, especially in the newly irrigated canal areas of the Punjab and in Sind, if certain long-deferred irrigation schemes could be carried out; but elsewhere it is a question of the relative value of different crops, and the native is slow to change, even when the relative prices of the two crops are changing in such a way as to alter completely the relative profit of the two crops. Thus the recent rise in the price of cotton must have changed the conditions in many cases;

but the uncertainty as to the future course of prices checked any tendency to substitution on a large scale.

The greatest difficulty, however, in India is the improvement of the quality of the crop, both in respect of the length of the staple and its condition as it comes to market. There is no doubt that India can grow much better cotton than the bulk of her present crop. Already something like half a million bales of fairly good staple cotton, much of it slightly over an inch, are to-day being grown in various parts of India, especially in Southern Madras, in the Dharwar District of Southern Bombay, in the Broach and Surat Districts of North Bombay, in Sind and in the Punjab, as well as small quantities in some of the native states; but the great difficulty is to maintain the purity of the improved varieties against mixing in the ginneries, with the consequent rapid deterioration of the improved seed, and also to get the cotton properly ginned and sent to market in a reasonably clean state. It is the latter feature of the crop which tells most strongly against the wider use of Indian cotton in Lancashire. In fact, one may say that the whole of the practical problems of improvement of the Indian crop centre in the ginnery, for the system of handling the crop throughout is in most provinces centred in small native dealers who, in many cases, buy the unginned cotton and gin it in their own factories, where also they mix it and sell the proceeds, both lint and seed.

This involves several undesirable consequences. These small dealers care very little for staple cotton, for which they have no market. All they care for is a heavy ginning out-turn, *i.e.* a heavy percentage of lint out of the seed cotton; and if they do find that a certain type of improved longer staple cotton is fetching a premium in the market, nothing will prevent them from trying to make up the quantity of the improved variety by mixing in some short staple with it. The result is, that the seed next year is mixed, and the improved variety at once deteriorates.

To remedy such a state of affairs is a very difficult and complicated business, which cannot be tackled unless the Government is prepared to step in and assist the trade and the growers to alter the conditions. But hitherto the Government have not seen fit to adopt compulsion in any way towards reformed methods; the trade will do nothing of its own initiative, and the planters can do nothing against the apathy of the trade which will not pay an adequate premium for good staple cotton or cotton in good condition. In the meantime the lack of the necessary conditions of a good ginning and marketing system is largely frustrating the efforts of the agricultural departments to produce and spread the use of improved varieties. Only in certain districts has real progress been made, and in one outstanding case, the Central Provinces, the variety found best suited to the district is unfortunately a short staple variety which on account of its very high yield and ginning out-turn entirely outbids all longer stapled but less prolific rivals.

One is, therefore, forced to the conclusion that under existing conditions any improvement of the Indian crop as a whole is bound to be slow, and that it would be foolish to expect from India alone the million bales annual increase which we are seeking. It is not that India *could* not do it, but simply that the inertia of a huge system, scattered over the whole length of a country that is really a continent of many different races and of endless differences of climate and growing conditions, is too great to be overcome in a short time. That is unfortunate, because time is of the essence of the matter to the world at present. There are other areas which can give us the increase some day; but none where it can be looked for within the immediate future. If India will not make the effort there is nothing for it but to wait for the others, and to do everything possible to push on their development. Thus in Africa there is still one area to be considered which might in course of time give us a substantial crop, namely, Nigeria. There, especially in Lagos and Northern Nigeria, the work of the British Cotton Growing Association has certainly proved the possibilities of great development, but until recently the difficulty was to find an improved variety which would be sufficiently superior in yield and value to the indigenous types to tempt the natives to grow it on a large scale. For cotton has been grown in West Africa probably from time immemorial, and the natives are fairly expert in the production of their own coarse cloths from the local growth. Now, however, a new variety has been produced which yields better than the old types, and gives a better ginning out-turn so that the association can offer a higher price for it, which is enough to turn the scale in its favour, and during the past season it is said that large quantities of this new variety were being grown and brought into the association's ginneries. It is perhaps still too soon to say that the corner has definitely been turned and that Nigeria is now on the high road to the development of a really large cotton crop for export, but there is at least reason to hope so. Even if it is so, however, there will still be need for very considerable expenditure in the way of opening up communications to facilitate the transfer of the crop to the coast; and it will still be a good many years before the crop can hope to reach even 100,000 bales.

There is, then, nothing else for it but patient, determined and continuous effort to encourage improved methods in existing fields such as India, and to push on as rapidly as is humanly possible the development of all possible new districts. The British Empire must face the question in a very different spirit from that in which the problem has hitherto been tackled, if any really adequate progress is to be made within the time available. The cotton industry cannot wait indefinitely for its increased supplies. Every year's delay means high prices, and though we have got accustomed in these days to high prices for everything we cannot with equanimity accept such a state of affairs as permanent. We can grow all the cotton we want in our own Empire if we will tackle the question seriously, energetically and *at once.* It will take millions of money to do it, and a good many years of time, but the sooner it is begun the better; and there is this further consolation: that every penny spent on the development of the cotton-fields of the Empire will also be assisting the development of our Empire in respect of other crops at the same time, and will be the surest and quickest way to improve the conditions of our colonies in every way.

This is no case of wishing to exploit the resources of our Empire for the benefit of English industries. On the contrary, the cotton-growing countries will find that at anything like recent prices, cotton will pay them exceedingly well, and probably much better than any other crop they could grow. We need, therefore, have no hesitation in pressing the matter on them, knowing that in this case the interests of all parts of the Empire are identical. An abundant supply of Empire-grown cotton means not only more cotton for Lancashire, but also the rapid disappearance of poverty from India, the development of our enormous territories in Africa, and the steady increase at reasonable prices of the supply of cotton goods upon which the people of all our great dependencies spend a large part of their free earnings, and which, indeed, represent to them the progress of civilisation. But the initiative, the driving power, and the capital must come from the Mother Country. In these days of promised reconstruction there is much talk of developing the resources of the Empire. Here is a clear case in point.

P.S.—The developments in the cotton situation since the foregoing was written have been so striking as to call for special notice. America did come into the war, with the result that there was a very strong movement in favour of the increased cultivation of cereals in the South as well as elsewhere, and corn prices went very high. At the same time the Government placed large orders for heavy cotton goods for the army.

At first the 1917 crop in America had the worst of luck, and the area sown (see Table F), instead of being a record, fell distinctly below that of 1916. This was partly due to bad weather during the normal sowing season, but also to a further reduction of the labour supply in the Cotton Belt through heavy drafts of negro labour to the industrial North. The advance of the boll-weevil also caused a reduction of acreage in Alabama and the south-eastern States. The bad weather caused a certain abandonment of acreage, and resulted in the worst figures on record for the first three crop condition Reports.

In the meantime demand had shown no signs of a real check, which would have eased the pressure on the supply. Though the consumption in England was necessarily affected by lack of labour and shipping difficulties, that of America, India and Japan, as well as neutral countries, was apparently still increasing, and it became apparent that the rise in prices was not sufficient to cause any material restriction of consumption. The world had begun to realise that cotton was still the cheapest textile available, because, as indicated in the text, cotton prices had not risen in proportion to those of other commodities in general and other textiles in particular.

The unrestricted U-boat warfare which began in February 1917 inevitably affected the imports of cotton, and early in June the cotton market in Lancashire suddenly awoke to the fact that the Liverpool stock was dwindling rapidly, that there was a serious shortage of the raw material in this country, and probably not enough to carry us through to the end of the season. The result was a scramble for the available stock, which in a few days forced prices up to levels out of all comparison with those quoted in the text. On June 20th the "futures" market of the Liverpool Cotton Exchange was closed, practically under Government order, with the spot price of American Middling at 19·39*d.*, from which it rose to 20·15*d.* in August. The Government appointed a Board of Control to carry the trade through the immediate crisis, which it is now clear will not be a matter of one season only, for unless a miracle happens the American crop this year will again be seriously short of the potential demand, and prices there are as high as in Liverpool—allowing for heavy freight and insurance charges. The Control Board introduced on September 10th a compulsory restriction of output over practically the whole trade. Another Committee has been appointed by the Board of Trade to go into the whole question of the Empire's cotton supplies.

After being practically closed for nearly four months the Liverpool futures market was reopened on October 1st, with an entirely new form of contract based on "Good Middling" instead of "Middling," and at the same time a system of official fixing of spot values was adopted.

During September owing to improving crop reports prices fell as low as 16·90*d.*, but recently a new rise to over 20*d.* has again taken place as the result of further fears of damage to the crop.

A new and alarming factor in the supply of fine cotton is the recently announced intention of the Egyptian Government to restrict cotton-growing in 1918 to one third of the cultivable area—*i. e.* about 1,350,000 feddans (acres), or a reduction of about 20 per cent. on the 1917 acreage (see Table G).

The statistics in the various tables have as far as possible been brought down to date (October 10th, 1917).

APPENDIX

TABLE A.—Statistics of the Cotton Trade in the United Kingdom

(*From the Census of Production*, 1907, etc.)

1 *Gross Output*—Value . . . . . . £	174,601,000	
2 *Materials used*—Cost	128,697,000	= 74 % of 1
3 Allowing for duplication the Gross Output is estimated at . . . . . . £	132,000,000	
4 *Output of yarn* (single) estimated . lbs.	1,800,000,000	
5 Valued at approximately . . . . . £	90,000,000	= 68 % of 3
6 *Consumption of Raw Cotton* of all kinds in Great Britain. Calculated from the Statistics published by the International Cotton Federation. Mean of Seasons 1906–7 and 1907–8 4,455,000 running bales averaging say 500 lbs. . . . . . . = lbs.	2,227,000,000	
7 Valued at approximately 7*d*. per lb. Based on the Season's average prices for 1906–7 and 1907–8 of American Middling, F.G.F., Brown Egyptian and No. 1 Fine Oomra, Indian . =£	64,970,000	= 72 % of 5
8 *Exports* of yarn . . . . lbs.	241,077,000	= 13·4 % of 4
9 Valued at . . . . £	15,417,000	= 17·2 % of 5
10 *Piece goods* made from the yarns retained in the United Kingdom (86·6 % of total)[1] . . . . . . yds.	7,087,680,000	
11 Valued (including cost of bleaching, dyeing, finishing, etc.) at . . . . . £	94,500,000	
12 Exports of Piece goods . . . . . yds.	6,297,708,000	= 88·9 % of 10
13 Valued at . . . . £	81,049,000	= 86 % of 11
Thus about 90 % in bulk of the original output of yarn was finally exported, *i. e.* 13·4 % as yarn and 88·9 × 86·6= 77 % as piece goods.		
14 Total value of Exports of Cotton Manufactures of all kinds (except hosiery, lace and ropes) . . . . £ or about 80 % in value of the estimated gross output of £132,000,000	105,043,000	

[1] Less yarns used for other purposes, *e.g.* sewing cotton.

TABLE B.—Position of Cotton in our Foreign Trade

(From the Statistical Abstracts, etc.)

In millions of £'s

Year.	Total: Imports.	Total: Re-exports.	Total: Net Imports.	Raw Cotton: Total Imports.	Raw Cotton: %	Raw Cotton: Re-exports.	Raw Cotton: %	Raw Cotton: Imports. Net	Raw Cotton: %	Exports of U.K. Produce: Total.	Exports of U.K. Produce: Cotton Goods.	Exports of U.K. Produce: %
1907	646	92	554	70	*14*	10	*14*	60	*14*	426	110	*26*
1908	593	80	513	56	*9*	8	*10*	48	*9*	377	95	*25*
1909	625	91	533	60	*10*	8	*9*	52	*10*	378	93	*25*
1910	678	104	574	72	*13*	10	*10*	62	*14*	430	106	*25*
1911	680	103	577	71	*12*	11	*11*	60	*12*	454	120	*26*
1912	745	112	633	80	*13*	11	*10*	69	*14*	487	122	*25*
1913	769	110	659	71	*9*	9	*8*	62	*9*	525	127	*24*
1914	697	95	602	55	*8*	7	*7*	48	*8*	431	90	*21*
1915	852	99	753	65	*8*	10	*10*	55	*7*	385	75	*19*
1916	949	98	851	85	*9*	10	*10*	75	*9*	507	104	*21*

TABLE C.—Spindleage and Consumption of the World's Cotton Industry in 1912–13

(Calculated from the Statistics of the International Cotton Federation)

Country.	Spindles.	% of World's Total.	Consumption in running bales.	% of World's Total.	Per 1000 Spindles.
					Bales.
Great Britain	55,653,000	*39·0*	4,274,000	*18·6*	77
India . . .	6,084,000	*4·2*	2,177,000	*9·5*	358
Canada . .	855,000	*0·6*	113,000	*0·5*	133
Total British Empire .	62,592,000	*43·8*	6,564,000	*28·6*	105
Total World .	143,453,000		22,932,000		156

TABLE D.—THE WORLD'S COTTON CROPS, 1902–1918
(*In Bales of approximately* 500 *lbs.*)

Season.	America.	India.[1]	Egypt.	Russia.	China.	Others.	Total.
1902–1903	10,758	3,367	1,168	342	1,200	801	17,636
1903–1904	10,124	3,161	1,302	477	1,200	751	17,015
1904–1905	13,557	3,791	1,263	536	756	803	20,706
1905–1906	11,320	3,416	1,192	604	788	936	18,256
1906–1907	13,551	4,934	1,390	759	806	1,027	22,467
1907–1908	11,582	3,122	1,447	664	875	950	18,640
1908–1909	13,829	3,692	1,150	698	*1,250*	969	*21,588*
1909–1910	10,651	4,718	1,000	686	*1,750*	950	19,755
1910–1911	12,132	3,853	1,515	895	*1,800*	967	*21,162*
1911–1912	16,043	3,288	1,485	875	*1,800*	1,058	*24,549*
1912–1913	14,129	4,395	1,507	911	*1,800*	1,171	*23,913*
1913–1914	14,610	5,201	1,537	1,015	*2,000*	1,340	*25,703*
1914–1915	15,067	5,209	1,298	1,247	*2,000*	*1,300*	*26,121*
1915–1916	12,953	3,738	961	1,465	*1,750*	*1,100*	*22,048*
1916–1917	12,976	4,273	1,022	*1,200*	*2,000*	*1,200*	*22,671*
1917–1918	*13,500*	*4,600*	*1,200*	*1,000*	*2,000*	*1,200*	*23,500*

The figures in *italics* are estimates only, very little statistical information or none at all being available.

[1] These are the Government estimates, which are notoriously understated. On the other hand, Indian bales are only 400 lbs. weight. Taking the Government figures as 500-lb. bales, as is done here, probably offsets the under-estimation fairly well on the whole, though roughly. The following table from the *Economist* of September 22, 1917, seems to justify this approximation.

Season.	Government Estimate.	Approximate Actual Crop.	
	400-lb. Bales.	400-lb. Bales.	500-lb. Bales.
1908–1909	3,691	4,744	3,795
1909–1910	4,718	5,341	4,273
1910–1911	3,853	4,974	3,979
1911–1912	3,288	4,643	3,714
1912–1913	4,610	5,019	4,015
1913–1914	5,065	6,684	5,347
1914–1915	5,209	5,279	4,223
1915–1916	3,738	5,407	4,326
1916–1917	4,273	4,818	3,854
Average	4,272		4,170

} Based on Exports, Indian Mill Statistics, and estimated domestic consumption.

TABLE E.—BALANCE OF PRODUCTION AND CONSUMPTION OF COTTON, 1904–1917

	World's Commercial Crops and Mill Consumption.[1]				[2] American Crop and World's Consumption thereof.			
	Mean Crops.	Mean Consumption.	Balance.	Average Price of American, Indian, and Egyptian.	Commercial Crops.	Consumption.	Balance.	Average Price of American.
1904–1905 .	19,648	17,726	+ 1,922	5·66	13,656[3]	12,664[3]	+ 992[3]	4·93
1905–1906 .	17,266	18,214	− 948	6·73	11,443	12,081[4]	− 638	5·94
1906–1907 .	20,815	19,523	+ 1,292	7·21	13,735	13,203[4]	+ 532	6·38
1907–1908 .	17,564	19,393	− 1,829	6·68	11,456	12,112	− 656	6·19
1908–1909 .	20,229	19,828	+ 401	6·29	13,831	13,157	+ 674	5·50
1909–1910 .	17,216	19,148	− 1,932	9·10	10,592	11,754	− 1,162	7·86
1910–1911 .	18,854	20,222	− 1,368	8·54	11,986	12,054	− 68	7·84
1911–1912 .	22,157	21,495	+ 662	7·09	16,108	14,515	+ 1,593	6·09
1912–1913 .	21,503	22,302	− 799	7·57	14,106	14,715	− 609	6·76
1913–1914 .	23,309	22,296	+ 1,013	7·52	14,882	14,541	+ 341	7·26
1914–1915 .	Complete statistics not available.				15,108	13,834	+ 1,274	5·22
1915–1916 .					12,938	14,812	− 1,874	7·51
1916–1917 .					12,940	14,046	− 1,102	12·33
1917–1918 .					***13,500***	***14,000***	− ***500***	—

[1] For sources and details see *The World's Cotton Crops* (by the writer), Statistical Appendix B.
[2] Figures of crop and consumption compiled by H. G. Hester, Secretary of the New Orleans Cotton Exchange.
[3] From Shepperson's Cotton Facts.
[4] Takings, not consumption.

TABLE F.—AMERICAN COTTON CROP STATISTICS, 1889–1918

Season.	Acreage Census. (Bureau Figures.)	Crop. (*Chronicle's* Figures.)	Yield per Acre.	Average Price of Middling
				Pence per lb.
1889–1890	20,175,270	7,313,726	0·36	6·19
1890–1891	20,509,853	8,655,518	0·42	5·00
1891–1892	20,714,937	9,038,707	0·44	4·12
1892–1893	18,067,924	6,717,142	0·37	4·61
1893–1894	19,525,000	7,527,211	0·39	4·19
1894–1895	23,687,950	9,892,766	0·42	3·44
1895–1896	20,184,808	7,162,473	0·35	4·38
1896–1897	23,273,209	8,714,011	0·37	4·22
1897–1898	24,319,584	11,180,960	0·46	3·47
1898–1899	24,967,295	11,235,383	0·45	3·28
1899–1900	24,275,101	9,439,559	0·39	4·87
1900–1901	25,758,139	10,425,141	0·41	5·16
1901–1902	27,220,414	10,701,453	0·39	4·78
1902–1903	27,114,103	10,758,326	0·40	5·46
1903–1904	28,016,893	10,123,686	0·36	6·94
1904–1905	30,053,739	13,556,841	0·45	4·91
1905–1906	26,117,153	11,319,860	0·43	5·95
1906–1907	31,374,000	13,550,760	0·43	6·38
1907–1908	31,311,000	11,581,829	0·37	6·19
1908–1909	32,444,000	13,828,846	0·43	5·50
1909–1910	32,044,000	10,650,961	0·33	7·86
1910–1911	32,403,000	12,132,332	0·37	7·84
1911–1912	36,045,000	16,043,316	0·45	6·09
1912–1913	34,283,000	14,128,902	0·41	6·76
1913–1914	37,458,000	14,609,968	0·39	7·26
1914–1915	37,406,000	15,067,247	0·39	5·22
1915–1916	32,107,000	12,953,430	0·40	7·51
1916–1917	36,052,000	12,976,000	0·36	12·33
1917–1918	34,600,000	***13,500,000***	***0·39***	—

TABLE G.—ACREAGE, YIELD AND PRICES OF THE WORLD'S CHIEF CROPS, 1913–18

Season.	Acreage.	Per cent. on 1913.	Crop.	Yield per Acre.	Liverpool Prices (pence per lb.). Lowest.	Highest.	Average.
American.	Acres.		Bales—500 lbs. approximately.	Bales.		Middling	
1913–14	37,458,000	—	14,609,968	·39	6·20	7·96	7·26
1914–15	37,406,000	100	15,067,247	·40	4·25	6·50	5·22
1915–16	32,107,000	86	12,953,450	·40	5·34	8·74	7·51
1916–17	36,052,000	96	12,976,000	·36	8·12	19·45	12·33
1917–18	34,600,000	92	*18,500,000*	*·39*	—	—	—
Indian.	Acres.		Bales—400 lbs.	lbs.		No. 1 Fine Oomra.	
1913–14	25,020,000	—	5,065,000	81	4·70	6·56	5·87
1914–15	24,595,000	98	5,209,000	85	3·94	5·00	4·46
1915–16	17,746,000	71	3,738,000	84	4·75	6·90	6·09
1916–17	21,212,000	85	4,273,000	81	7·10	18·30	*11·00*
1917–18	*22,500,000*	*90*	*4,850,000*	*85*	—	—	—
Egyptian.	Feddans.		Kantars.	lbs.		F. G. F. Brown.	
1913–14	1,723,094	—	7,684,172	444	8·15	10·45	9·44
1914–15	1,755,270	102	6,490,221	369	6·30	8·30	7·34
1915–16	1,186,004	69	4,806,331	406	7·50	11·90	10·42
1916–17	1,655,512	96	5,111,080	310	11·60	31·50	21·56
1917–18	1,677,310	97	*6,000,000*	*360*	—	—	—
1918–19	*1,350,000*	*78*	*4,860,000*	*360*	—	—	—

The figures in *italics* are estimates.

TABLE H.—THE WORLD'S COTTON SUPPLY, AND THE BRITISH EMPIRE'S SHARE IN IT.

Grade.	Quality.	Stap Inches.	Where Grown.	Pre-war Prices, 10/7/14.	World's Crops, Bales.	Bale Weight.	Empire's Share	Per cent.
				Pence per lb.				
I.	Best Sea Island . . .	2 and over	Islands, South Carolina .	12½–18 [1]	10,000	400 lbs.		
			West Indies	13 –18⅓ [1]	5,000	400 ,,		
					15,000		5,000	33
II	Second Grade Sea Islands	1½–1¾	Florida and Georgia . .	11½–12½	70,000	400 ,,		
			West Indies	,,	2,000	400 ,,		
	Best Egyptian (Sakel, etc.)	,,	Egypt	10–11½	430,000	740 ,,		
					502,000		432,000	86
III.	Egyptian	1¼–1⅜	Egypt	8½–10¾	1,000,000	740 ,,		
			Sudan	,,	25,000	400 ,,		
	Staple American . .	,,	Mississippi Delta, etc. . .	up to 10½	200,000	500 ,,		
		1⅛–1¼	Nyasaland, Uganda, and East and South Africa .	6½–10	50,000	400 ,,		
		1 –1½	Peruvian	7—10	125,000	250 ,,		
					1,400,000		1,075,000	77
IV.	American	1 –1⅛	U.S.A.	Middling 7-33	15,000,000	510 gross		
			Mexico		150,000			
		¾–1¼	Brazil	6–9	400,000	250 ,,		
		⅞–1⅛	Russia	(5–7½)	1,000,000	250–500		
		1 –1⅓	West Africa	6¾–7½	15,000	400 gross		
		¾–1⅛	Levant	(5–7½)	100,000	440 ,,		
		1	India	up to 7d.	250,000	400 ,,		
		1	China and Corea . . .		250,000	500 ,,		
					17,165,000		265,000	1·6
V.	Indian	⅝ up	India	3¾ upwards	5,000,000	400 ,,		
		,,	Russia		400,000	250-500		
		,,	China	(4–6½)	1,800,000	500 gross		
					7,200,000		5,000,000	69
					26,282,000		6,777,000	26

[1] Crop lots up to 40*d*.

METALS AS THE BASE OF IMPERIAL STRENGTH

By OCTAVIUS CHARLES BEALE,
ex-President and Representative of the Associated Chambers of Manufactures of Australia

THE exploitation of the accessible deposits of metallic ores all over the planet proceeds in an ever-increasing ratio. Metals are not only essential to every branch of industry, but the diversification of the activities of man requires their production in greater variety as well as augmenting volume. With the well-grounded apprehension that the present rate of use may cause privation to a not remote posterity we have in our present consideration nothing to do, unless, indeed, it serve to sharpen our interest to secure the position of our own Empire.

I shall endeavour to show that metals are basic to Imperial strength, yet only so in common with other products as set forth in the several papers of this series. Still I may seek to apply a special emphasis to the conservation of our mineral resources, to an organised resistance to foreign exploitation of them, for the very reason that we are dependent for supply upon natural deposits as contrasted with agricultural products that may depend only upon the application of human energies to soil and climate.

Two years before the outbreak of the war, the President of the Trades Council of Newcastle, N. S. Wales, with his secretary, called upon me in Sydney to ask me to join them in the effort to establish the zinc-smelting industry in Australia at their great coal-producing port. Eager to do so, we consulted the Sydney Chamber of Manufactures, the Labour representatives addressing the members upon the subject. Naturally, support was accorded.

In pursuit of the object I drove to Broken Hill, 850 miles west from Sydney, so as to call at the Cobar copper-mining camp on the way and to view the country in the vicinity of Broken Hill. You are aware that this is the greatest zinc-producing field of the world, and it is certainly open to vast development, for there can be no mistake as to the metalliferous construction of the rock formations all around. My object was in chief to ascertain the views of, and probability of support from, the mining people themselves, more especially those who have the votes—the miners and townspeople. It is a Labour stronghold.

The Secretary of the Trades Hall, an old acquaintance, promptly and positively assured me that his people were entirely with us who advocated localising the smelting industry in Australia and thus wrenching it out of the hands and control of the Germans. There could also be no mistake as to the desires of the highly-trained, alert-minded men who direct the mining and concentrating operations. To that point, therefore, all was well.

Returning, our committee collected its statistical ammunition, and here arose a phenomenon that has puzzled me ever since. The very agency, or more correctly some of its officers, through whom the long contracts for the entire output of the concentrates had been made and through whose hands deliveries and shipments were effected, furnished us with accurate figures and genuinely assisted the movement. No trouble was spared and finally we came with our formal deputation to the Minister for Trade and Customs, then Mr. Frank Gwynne Tudor.

Up to that point—publicity and driving home the wedge—all had run smoothly and hopefully. Nothing could or will persuade us that the Minister's head and heart were not absolutely with us, but we found ourselves up against a stone wall. We were "turned down" and defeated. The administration would make it no part of their policy, by imposing export duties upon the metal contents of ores or concentrates or in any other form that could be suggested, to enforce the production of the basic metals in Australia, for our own use and for export.

To say that we who approached the subject solely from the national standpoint, in no way pecuniarily interested, but with strong convictions and clear foresight, were disappointed and angry, is to put the thing mildly. We manufacturers had during long years declared that without its own independent metal industries a country cannot defend itself against aggression, nor can it become a nation. After that defeat I have myself declared in public meetings before the war, that those metals would return to our own sons' hearts in the shape of bullets.

Secretary of State Michaelis, Minister-President of

the German Department for Grain Supply, declared a few weeks ago that the Germans will find war to be a stern schoolmaster, both during the conflict and after concluded peace. So do, and shall, we British find it, one of the lessons of this rude monitor being that without economic independence you cannot have national independence and security.

Five hundred thousand tons of concentrates containing rather more than one-half basic metals, lead, zinc and copper, as well as some gold and much silver, were annually shipped to Germany in German steamers. The shipowners were correlated with electrical companies and supplies of machinery, tools, railway material and all other requisites. The Germans not only poached on John Bull's preserves, they barricaded him with assistance of his own laws and lawyers out of his own fields, the while they flipped him on the nose out of pure malice, knowing he would take it, not as any lion but like a lamb.

The servility of our public men towards the Germans, whom undeniably the bulk of our British people so deeply distrusted, was as astonishing as it was dangerous. At the Kaiser's birthday banquet in Sydney, just before the war, presided over by the German Consul-General who was at the time openly organising all German settlers, urging them to maintain their language, schools and nationality even if and whilst remaining naturalised British subjects—at that banquet the President of our Interstate Commission declared that we Australians ought to be grateful to the glorious German nation for doing so much business with us, as set forth by our exports and imports!

And the British Minister for War, in London, on December 9, 1911, in his sixth year of control of the Empire's defence, used these words: "The thing I most desire is to see Germany expand herself overseas, and make her beneficent influence, that of a great and civilised nation, felt in the distant parts of the world." Of course he meant India, Australasia and China. We felt it sure enough in Australia, New Zealand, New Guinea, Samoa and the Pacific generally, but to satisfy the doctrinaire in power we did not feel it half enough. Yet your brethren in the Pacific, who have given and will continue to give proofs of their love, and of their convictions, deeply resented the spread of German influence, military and commercial, in the mightiest ocean of the planet.

Moreover their indignant protests fell upon the deaf ears of those British statesmen who were so concerned in the furthering of German expansion. The same statesman, the most prominent exponent of the policy of the day, said at Dalston in June 1912: "The German Emperor is something more than an Emperor—he is a man, and a great man. He is gifted by the gods with the highest gift that they can give—I use a German word to express it—Geist. He has got Geist in the highest degree." Such phraseology cannot be applied to any Englishman that has ever lived, for the attribution is due only to one Personality that the world has known—and that was not Wilhelm der Zweite.

We may smile at this extravagance of sycophancy and deem it harmless, so that it may be forgotten. But history will not have it so. Our War Lord on January 15, 1914, at Holborn said: "Europe is an armed camp, but an armed camp in which peace not only prevails, but in which indications are that there is a far greater prospect of peace than ever before. No one wants war!" And at Highfield, one week before the beginning of the cataclysm, "I am not in the least afraid of the invasion of German armies."

Could we attribute such sentiments and conclusions to mere predilection there were no more to say. They might remain amongst the bygones. But we know for certain that at least in mercantile circles there exists amongst some of the traders an undisguised resolve to re-open connection with our foes at the earliest possible opportunity. We have been told directly, not as of hearsay, that merchants are retaining certain agencies for German manufactures, "because," say they, "if we do not hold to them, our competitors will rush for them." The same with regard to supplies, from our overseas domains, of manufacturing material.

Now is the time to decide upon one of two practices —I will not say policies—open to us when our energies shall be permitted to return to the normal channels of industry. We can re-admit our sworn enemies to our marts and clubs and homes, even if we stop short of placing some in Parliament, or we can proffer our products and potentialities to our friends who have fought for the common liberty and honour and hope. We have left behind us the old fallacy that "values are arrived at by the higgling of the market." Unless we want a renewal of the "bath of blood" we shall arrive at values otherwise, we shall renew and perpetuate the Grand Alliance by voluntary commercial preference devoid of higgling. To say that it cannot be done and that the scheme is Utopian can only be supported by sophistry, for the practice is one of daily existence. Our great Oversea Dominions give preference to the Mother Country, and in some instances to one another. Now is the day of salvation, because if we do not arrange a sincere and generous bond, free from all bargaining, with our faithful Allies whilst in the throes of the fight, it may be all too late when discussions for peace shall commence. First and foremost there should be substantial preference in prices and deliveries of manufacturing material. That is the head and front of what I have to lay before you to-day.

The immense power of production, in our Empire, of metals and minerals that are basic to all manufactures and to the enlargement of the production itself, must be laid open preferentially to our own States and to the Allies. It is for the several Parliaments, or for the appointed governing bodies in the case of Crown colonies, to decide upon the methods of preference. Again we must point to the principle of present action, which only requires extension, for it would be presumptuous in any person or association to declare how far or in what manner any State or States can confer preference in relation to one or more of their products.

A very fine example was set by that jewel of the Pacific, New Zealand. In devising their Tariff many years back a handicap of double duty was placed upon cement. That was fully intended to hit German mercantile, maritime and manufacturing interests, in favour of Great Britain, all in one blow. We all knew that. One moment to explain. Cement is not always ordered, it is often consigned at a venture. It is useful "stiffening" for ships so that they may carry light goods. With it also shipowners can make up full cargoes and get a good price for freight on the other lines. The eyes of the New Zealand legislators were wide open, no opiate films of fetish worship, so they acted upon sound, manly sense, commonly called horse sense. They did many other wise things, having their goal in view and—*ex uno disce omnes*.

Copper

When the war broke out the position of the copper industry in Australia, and indeed in the British Empire, was a pitiful humiliation and never to be forgotten. The whole output was under foreign domination and tight control. Years before the war I have asked the largest Australian producers why they did not sell directly to consumers so as to preserve a kind of independence. That was from the national standpoint and with a desire to know the causes. It was very courteously explained that such dealing was entirely impracticable, commercially impossible. We were shown that the world's output was controlled from New York by a hyphenated " syndicate " consisting of a mysterious but omnipotent five, and that any producers attempting to set up direct trade would be inevitably swamped by the unseen but very real power. You all know what a long time it took in our own courts to blow aside the smoke-clouds and to tear away some of the barbed wire entanglements, so as to expose the citadel of the militant five. The fortress was indeed exposed, but that it has been captured to the freedom of the nations no one will attempt to assert. It has been believed that the central capitalistic power resided in the Rothschilds' Frankfurt bank, but there will never be means of proof.

When the great explosion choked up the old channels and destroyed means of communication, the price of copper fell wofully, though in flattest contradiction to all previous experience and to rational expectation at the outbreak of war. The Australian banks would advance no more than £40 per ton, for the ostensible value was only £45. A sudden closing of the great mines was dreaded before it would be possible to arrange for direct sale to the principal consumers.

The Electrolytic Smelting Company of Australia with works at Port Kembla, which I have visited upon different occasions, is the largest producer in the British Empire. It was indeed, up to the war and probably is still, the largest outside of the United States, the capacity being 29,000 tons electrolytic and 12,000 tons fire-refined copper per annum. Yet Australia drew no pipe, nor wire, nor was a sheet rolled for her own large consumption. From year to year we manufacturers reproached our Government bitterly enough about it. No, not bitterly enough, for no attention was paid. We said, again and again, that as we did not roll a sheet of brass there could be no production of cartridges for defence if trouble should come. Without local manufacture of basic metals, iron, steel, brass and the rest, we could not be a nation.

But I have myself inspected the vast works of the Allgemeine Elektrizitaets Gesellschaft at Berlin, who used a large proportion of our Australian electrolytic copper, as preferring it to all other. The purity was 99·99 per cent., being absolutely clear of lead. Much or most of the remainder was worked up by the Imperial German munition factories. It all passed through London, through Aron Hirsch und Sohn, about as British as the recently retired Ignaz Nathan Tribitsch (alias Lincoln). You may remember that when the German agent was found a seat in the House of Commons, the then Minister for Foreign Affairs hastened to telegraph to Herr Tribitsch, "I congratulate you upon your splendid win!" The German had defeated a good Englishman for Darlington, so that it was, to be sure, a splendid "victory" in the old home of the Stephensons, a centre of the metal industry of England on that glorious north-east coast that is the chief strength of the British Navy and a very citadel of Imperial defence. You see no connection between the cash of your quondam foreign rulers in the metal domain and German interference in British Imperial politics? Are you so very sure?

The Australian Government is fully awake now. The family gaps, the empty chairs, the blinded and crippled sons, will be adequate reminders that only the strong man armed keepeth his house, and that all that is worth having is worth protecting. And I make bold to say of the great Australian-owned mines and refining companies that their directors and managers are within my own knowledge solid and strenuous in their practical patriotism. A very recent and valuable proof of it is that in connection with Port Kembla a branch company has provided a capital of £200,000 to turn out all usual copper parts for manufacturing purposes: wire, sheet, tubing, cables, brass, bronzes and other requisites. Such an installation emancipates the Commonwealth,

New Zealand and adjacent possessions from external interference or control in an important department of industry.

LEAD

The largest individual silver-lead smelting works in the world are at Port Pirie in South Australia, connected by rail with Broken Hill, the latter being also the largest mining camp of its kind in the world. It is really refreshing, not to say exhilarating, that for a number of things of basic importance, we can claim for the British Empire and in particular for Australia—that we lead the world. But if these essentials were owned by Germany, or owned by Americans, your present claim would be only an empty boast, paltry and contemptible. No pains should be spared to retain possession and control within the Empire.

The annual capacity of these works is from 150,000 to 200,000 tons of pig lead and from five to six million ounces of silver. The Company produces litharge, antimonial lead, spelter and other manufacturing material, possessing its own coke works, flux quarries and the like. "Its ordinary soft pig lead stands exceedingly high in the estimation of consumers, for chemical work and for corroding. Assays show an average of lead 99·988 per cent." (Quin.)

SPELTER

We have seen how the Germans held the whole zinc output of Australia, as also of other countries in their absolute control, disposing of the product as they thought fit—an autocracy. You have also before your minds the monstrous decision of a British judge, who declared from his exalted position that a German Rubber Company formed with German capital, consisting of German shareholders, ostensibly a branch of a company domiciled in Germany, but trading in England in war-time, was still an English company, and not German, because it registered in England its name, articles and occupation! Can casuistry further go? But in contemplating the metal position we have to keep such a decision in mind. In war-time the Government itself is forced to move by public clamour, whereas in peace-time there is no clamour, and, by consequence, no movement. The reversal of his shameful decision does not therefore secure our future.

Forced, moreover, by war to face the truth, the zinc position has been rescued by our own citizens. A Zinc Producers' Association has been formed to handle the entire output of Australian concentrates. All the producers are members. While the war lasts, a quantity of at least 100,000 tons annually of concentrates will be treated in the United Kingdom. For ten years after the war, Great Britain will take the same minimum annually; 40 per cent. of the zinc concentrates of Australia will be treated in that country, and 45,000 tons of spelter will be purchased thence by Great Britain. To treat a portion of the concentrates reserved for Australia the Electrolytic Zinc Company of Australia, with a capital of £1,000,000, has been formed, to operate near Hobart in Tasmania. Any excess of concentrates which cannot be handled locally or in Great Britain will be treated in Allied countries. There are other large zinc-smelting works springing into existence in Australia. And that there are vast deposits of ores, unworked, both in the south and the north of the Continent, can be demonstrated. (Quin's *Metal Handbook and Statistics*, 1917.)

In the zinc department Broken Hill is the most important field in the British Empire. The concentrates contain about 46 per cent. zinc, 8 per cent. lead, and 15 oz. silver per ton.

Now, although the ore production of Germany had been steadily declining, her zinc production had increased from 190,000 tons in 1904 to 280,000 tons in 1913. That compared with England puts Germany first by five times the output. And such German output depended upon British Australian ores as principal source, whilst English production for four or five years showed a declining tendency. It was due to *laissez-faire* pure and simple. Do let us put an end to the awful falsehoods that it was attributable to the accursed drink, to the British workman, his strikes, and his lack of skill, to defective technology, to all but the real thing. It was due to false doctrine and wicked neglect. I am painfully tempted to declare the names and quote the doctrines and practice, but have done so abundantly in former writings.

War did not suddenly correct our technology, make us all abstainers, or qualify our workers. No, it dragged away our best men in all branches of industry, increasing our difficulties to distraction. But we remain the same dram-drinking, meat-eating race that shifts this planet and that has taught Germany, amongst the rest, more of practical mechanics, metallurgy, chemistry and electricity than they have themselves acquired from any other source whatsoever. War made us look facts in the face, respect one another, help one another, turn back to our own national glories, cling to our own race and country as our "spiritual home." Hence we rescued the mining and metal industry as one of the bases of our power, as we are likewise determined to emancipate ourselves in other fields of activity.

We are full of hope, but have by no means thrown off the shackles that have been rivetted upon our limbs during the half-century of blindness under *laissez-faire*. As, for instance, out of the world's zinc-production, in 1912, of 975,000 tons, more than one-third (354,000 tons) was controlled by the German Syndicate and nearly two-thirds (571,000 tons) by the International Syndicate. Otherwise stated, out of 975 thousands, 925 were "controlled" by two swindlicates, as they are very irreverently called. Japan accounts for the small balance. Consortium, combine, cartel, syndicate, all are powerless when they come into conflict with a resolute Government, be the State great or small. To this point we shall recur.

The whole saving of zinc from millions of tons of tailings and from refractory ores, was due in no wise to the Germans, although they snatched and held the turnover and the chief profits. It was because of the curious and clever invention of the flotation process, doubtless familiar to you all.

Mr. H. K. Picard at the Institute of Metals said, "The Germans had nothing whatever to do with the technical development of the flotation process from first to last. I am able to say that, because I have been connected with it from its inception. It was an invention that was made and developed in all its details by British metallurgists, and it was only after they had produced the concentrates that arrangements were made with the German smelters to treat them."

Following the lamentable example of many persons in high places in England, Australian mining folk allowed themselves to be bluffed by German effrontery. It was successfully dinned into their ears that Germany had a specialty in retorts which human ability could not produce elsewhere. Forgive me if I weary you with illustrations, but we are all the time driving at practical conclusions.

Zinc concentrates are treated, distilled, in furnaces, at very high temperatures both in Germany and Belgium. The method was not used in America, the clay being—it is alleged—inferior. The process is known as "hard driving" and is extremely severe upon the retorts. Undoubtedly a special fire-clay is essential. The current belief, sedulously fostered by the vigorous assertions of the Germans, was that no such clay could be found in Australia. Now, relative to anything in the range of metals and minerals, that was an outrageous assumption. However, the manager of construction, employed by the Broken Hill Proprietary, a keen and canny Scot of the kind—Heaven be praised!—that it is unsafe to bluff, told me the following—

"We were assured too often that fire-clay could not be obtained suitable to the purpose, and we were shown failures of various kinds. But we sent out 'prospectors' to discover it. They brought in, after some search, specimens of a sort that gave them sanguine expectations. To come to the end of the story, we had bricks made of it, put a German brick on top of an Australian, and subjected both to extreme temperatures in a furnace until the German specimen finally fused over our own. We bought the land containing the clay, and thus ended the controversy."

There were usually shipped from Australia from 450,000 to 500,000 tons of concentrates per annum, in German steamers as you would expect, the office of the exporting agent being in a building of the Norddeutscher Lloyd—just as it happened. I have been there in the course of my inquiries. The German consortium, or conspiration, could send prices up or down at their discretion and as the production could be made unprofitable, within discreet limits. And the German steamship lines must be considered. Perhaps out of sweet shamefacedness a twenty-fifth part, 20,000 tons, of this British product was allowed to go to the United Kingdom. John Bull was becoming accustomed to a crumb from the table of Dives. You know the story of the final cancellation of the long-term contracts, so designed with cunning and foresight that war itself was not to annul them.

It may be added that whilst the production of zinc ores in Belgium, France and Germany has been steadily declining, there are grounds for the belief that Australia is inexhaustible and that Canada—in particular British Columbia—will henceforth be a large producer.

There is strong reason to believe that the metal world will get along in future quite happily without German intervention. I know that in Australia they say, with a reinforcing adjective, they will do " a sight better."

Tin

The story of tin may soon be told, as to results. But the internal history of metal exploitation, if it could be laid bare, would be a fascinating revelation.

In 1914 the British Empire produced 76,000 tons out of the world's supply of 113,000 tons, more than twice as much as the rest of the planet. Of our portion the Straits Settlements yielded about eleven-twelfths. Such a gorgeous plum in the pudding was not overlooked by hyphenated Americans. Accordingly, some years ago, an "American" syndicate acquired land near Singapore with the announced purpose of smelting tin-ores, the produce of Malaya. It is more than half the battle when a would-be monopolist obtains the handling of the bulk of that which he proposes to control. Smaller fry can be leisurely swallowed afterwards. But the Government of Malaya were alert and *laisser-aller* had no attraction for them. They made such fiscal arrangements as vested in them the disposal of their country's product. The foreign syndicate was fain to sell back, at a loss, their land acquisition, and the smelting of all the tin ores remains in British hands, the article itself in British disposition.

Pray do not think it irrelevant if I add that before the war broke out, the same Government sent to the Mother Country a cheque for a cool three millions sterling to pay for the splendid battleship *Malaya*, I believe the first example of its kind. The ship has since taken a glorious part in the active defence of the Empire, and you know how the example has been followed. By way of parallel with the tin production it may be mentioned that of the world's supply of rubber 92 per cent. is of British providing, Malaya and the Straits Settlements being far in the lead. Thus is a truculent enemy made to feel what it means when, in the cause of civilisation, England brings on the whole puissance of her Empire. Further, from the financial point of view, the value of tin produced by us was, in 1900, £6,900,000 and, in 1913, £13,500,000. Doubled in a dozen years.

TUNGSTEN

Burma heads the list as the world's greatest producer. Only within the last few years this metal, its salts and alloys, have been brought into general use. " Its chief value lies in the fact that when added to steel the latter acquired the property of self-hardening. This renders it extremely valuable for high-speed cutting tools, and, in fact, four-fifths of the total output of tungsten is now absorbed in the manufacture of ferro-tungsten " for the purpose named.

But the whole of the British product, at least, was handled by a German company, the value of the Burma yield being, according to the *Handbook of India*, only about £150,000. It would be interesting to know what the value had swollen to by the time the Germans issued the hardened steel in saleable form. Only those who are intimately connected with manufacturing industries can estimate the importance of retaining control of such an essential, though the cash measure of it be small.

QUICKSILVER

Following in the same connection, attention ought to be frequently drawn to a monopoly which has for about half a century remained in the hands of a German firm of bankers. It is silent, but the control is understood to have been quite undisputed during that long period. The remedy is perhaps not so easy of suggestion and would require lengthier consideration than can here be given. Certainly we know that in Australia mines for working cinnabar, the usual ore of mercury, have repeatedly been opened. Spain and Italy are chief producers. Australia is not listed at all. It pays monopolists to buy out "claims" in that, as in other metals, and to leave them unworked. Even in America the phenomenon is far from rare. The question that arises is—would it not be well for a Government to acquire such mines and work them for the benefit of the country concerned?

Manganese

Of this, British India yields the greatest proportion and the out-turn could be reckoned at 800,000 tons a year, varying with demand and available freights. Again, this is a trump-card in the steel trade, for its destination, and even the price, can be controlled under ordinances.

Mica

India issues half the world's supply of this mineral, whilst of

Asbestos

the Canadian production has so vastly increased that she apparently can supply the whole needs of the world.

Nickel

Of this metal, whose beauty and usefulness are so familiar to all, the two great sources are Canada and New Caledonia. Here, again, comes in the control by a New York metal syndicate, the real exploitation being out of the hands of either British or French. The actual interest of France in the mines of New Caledonia must now be small indeed, but those great fundamental facts are just where statistics utterly fail us. They are hidden, carefully hidden in board-rooms and iron safes, and nothing but a Royal Commission of independent business men could extract the knowledge that ought to be in the full view of Parliament and public. Not as to nickel only, but in all departments of metal provenance, this obscurantism involves the bases of our national defence and prosperity.

Monopolists holding metals in their control have a strangle-hold upon a nation's life. Political Economy tells us nothing whatever of all that, nor is it in any way concerned. Its axioms, arguments and conclusions, are more unfortunate even than the phrase itself, forced upon the world by François Quesnay, who joyfully announced his discovery of the quadrature of the circle. But we can have, and we surely need, National Economy, *Volkswirthschaft*. To the political economist the value of the export is all that concerns him, for he takes no account of the ownership. The value appears in the statistics of the country, yet the country may have no concern with the value beyond the quarrymen's wages and cheapest transport to foreign steamers. Their argued conclusion is, that exports are always paid for by imports, whereas the whole value contents of archipelagoes have been exported from them and nothing at all imported. In like manner a country may be wholly or partially depleted of its metals and minerals by foreign exploitation, values being realised and revenues spent, elsewhere. If it be a free people it is up to its legislators to protect the national interests by simple regulation, whether fiscal or by other action in restraint of foreign avarice.

These are the words of the late Yves Guyot in his *Principles of Political Economy:* "When Quesnay, following de Gournay, repeated the formula *laissez-faire, laissez-passer,* it meant 'Respect natural law. Do nothing to disturb the natural order of the production and distribution of wealth.' It is the first formula of an art, which, renouncing *a priori* reasoning, has learnt to confine itself to the application of existing laws." Well, if there be any natural law it is expressed in the German proverb, *Wo Tauben sind, da fliegen Tauben zu* ("Where pigeons are, thither pigeons fly"): or the English "Money makes money." If that is what we want, then, let us stick to Political Economy—and the devil take the hindmost.

THE RARE METALS

Bismuth

Countries of production extremely few, Australia being one of them. Also subjected to a close monopoly by the owners of the world's quicksilver.

Thorium Group

The phrase much used in the defence of a monopoly, by Auer von Welsbach and his satellites, which was obtained by another way round. By a comprehensive claim for a patent of monopoly over the use of "rare earths," backed by overwhelming financial power, immense revenues were extracted from civilised mankind to the great retardation of progress in glow-lights. After the battle for monopoly had been won as against the harmless, useful public—that pays for everything and gives its wealth, limbs and lives when trouble occurs —I said to a winning barrister, "Surely those claims were very wide, covering a string of natural agents indefinitely." He replied, "I only wish I had been on the other side!"

Our patent laws are wrongly so-called, for the true expression is "monopolies of invention." A volume could be written upon the unrighteous exploitation of our people for the benefit chiefly of avaricious and idle foreigners. The public knows nothing of these exactions permitted in some countries and prohibited in others, because lawyers in Parliament are not much addicted to enlightening the people in such matters. The attempted monopoly, in legal form, of the use of the natural agent cyanogen in the extraction of gold, was one of the most notorious. Yet, in countries where the monopoly was permitted, the profits were great.

Molybdenum

Is also employed by manufacturers of incandescent electric lamps. As Australia is the chief source it is probable that care will be taken to prevent at least German or Austrian exploitation. That the total value is small may be only a point in favour of the monopolist in obtaining high profits out of his ultimate products.

Iron and Steel

By way of explaining our success in Empire building we have all heard *ad nauseam* the aphorism "The flag follows trade." It puts in the shadow the work and the exploits of our brave soldiers and hardy sailors, our bold explorers and self-denying missionaries, who have upheld the honour of our race, made stable the Imperial sway and spread over all the world the glamour of the British name. But back of these again were the industrials who made things possible and in their way, too, were enterprising.

In the early part of the eighteenth century, Abraham Darby, of Coalbrookdale, first smelted iron from the ore with coke, revolutionising the old practice of using charcoal. He and his son of the same name continued and improved the method. They erected the first iron bridge ever set up in England or elsewhere. I am told that it remains as a testimony to their foresight, skill and enterprise, having resisted for nearly two centuries the stress of traffic and the storms of time.

Thus the brown-coated Quaker, by his epoch-making discovery, was one of the founders of England's greatness and himself an empire-builder. He "wrought out a monument more lasting than bronze, loftier than the Pyramids' royal pile," but I fear one would have to search vainly for his statue.

The *Economist* of December 16 last gives the following figures—

Germany exported in 1900, 800,000 tons iron and steel.

United Kingdom exported in 1900, 3,300,000 tons iron and steel.

Germany exported in year before the war, 6,000,000 tons iron and steel.

United Kingdom exported in year before the war, 5,000,000 tons iron and steel.

From Quin's *Metal Handbook*, I find—

Iron, Germany produced in 1913, 30,000,000 tons.
  ,, United Kingdom produced in 1913, 10,000,000 tons.
Steel, Germany produced in 1913, 19,000,000 tons.
  ,, United Kingdom produced in 1913 under 8,000,000 tons.

America produced much more than either of them although the ore and the coal are separated in America by long distances, whereas in England and Scotland they are practically side by side.

To judge by the number and diversity of useful inventions that our nation handed to mankind, we must have been in deep necessity before the full blossoming of the "first principle of political economy," to which the disconcerting figures above quoted are attributable. But necessity is a prolific mother and we may also thank her for the progress of the steel industry, whereby at least 3,000,000 tons more according to the *Economist*, were produced last year than in the year before the war. I am assured, however, by men in the steel trade, that the increase during the war is really much greater than that I have quoted.

During a quarter of a century I have made it a rule to obtain and study German newspapers, as well as other of their literature, and until the war have never been a week without them. For years, too, I collected the many pamphlets of the Alldeutsch Verband and of sporadic authors setting forth, in imaginative history, the destruction of the British Empire. *Arbeiter und Flotte* (The Working-Class and the Navy); *England in deutscher Beleuchtung* (England in German Illumination), by Graf Reventlow, who has been conspicuous of late; *Die Abrechnung mit England* (Squaring Accounts with England); *Seestern* 1906, in which the conflagration that afterwards broke out at Sarajevo is made to start in 1906 in Samoa; *England und die Engländer*, and ever so many more. No sane Briton who read these longings of the German heart could for a moment doubt what was in preparation. Probably very few British statesmen read anything German, but some did and either their self-deception was amazing or else they grossly deceived their nation. History cannot fail to record it, for an explanation will be demanded, even by this generation. Permit me to translate a recent article from the *Frankfurter Zeitung*, signed by Prof. von Leyden :—

"When peace shall be re-established no self-respecting German can ever consent to live in any place where an Englishman is to be found. There is no possibility of compromise upon this point. We are bound to swear a national vendetta against the British, never to pause in our preparation for another war, never to neglect any effort until we have destroyed for eternity the last vestige of British power.

"The Russians must be boycotted almost to the same degree. They share with the abhorred English the responsibility for this war. They belong to the conspiracy which aims at destroying German Kultur and to undermine German power. They are Eastern barbarians just as the English are Western barbarians. All Russia must be banned from civilised society.

"As to the French, we do not perhaps cherish for them the same violent hatred, but they must share in the contempt that we feel for the British and Russian allies."

Thus are we completely forewarned by the implacable foe whose Punic faith and Punic ferocity we know so well.

This is not the last of wars nor perhaps the worst. *Si vis pacem para bellum.* "If you want peace prepare war," for the maxim is usually mistranslated. Without economic independence there can be no national independence. That is the slogan of the British Empire Producers' Association, to which I have the honour to belong. It describes also the motives and action of the Australian Associated Chambers of Manufacture, who have appointed me their spokesman in London. During a long series of years, without a single dissentient voice in their Councils, they advocated Fiscal Preference between the States of the Empire and have rejoiced in its partial attainment.

These Australian manufacturers employ 350,000 people and produce annually goods to the value of £166,000,000 sterling, constituting by far the largest producing interest of the Commonwealth. The respective figures for Canada are very much larger; and there, again, manufacture is of predominant importance. There can be no question of the great self-governing Dominions adhering to their policy, or more correctly speaking, the loyal practice of Preference within the Empire. But it is equally certain that they will not consent to an Imperial Customs Union on the basis of Free Trade. As each is bearing its own burden of defence and military preparation to secure the future—

begun long before the war—so each will make its own fiscal arrangements.

The suggestion now before men's minds is that "first-call," substantial preference, in some form to be decided by each Parliament separately, shall be given to our own people and to our fighting Allies. As partly shown herein, we are the principal producers of materials, wrongly called raw materials, for consumption and manufacture. Amongst those of first importance are some that our Empire produces solely. The power is immense, but hitherto we have neglected it, we have let it slide, we have permitted foreign exploitation with our eyes open, by yielding to a fatuous philosophy and a pusillanimous anxiety lest we displease the arrogant German.

THE WHEAT SUPPLIES OF THE BRITISH ISLES

By HUGH R. RATHBONE, M.A.,
Member of the Royal Commission on Wheat Supplies, Member of the Mersey Docks and Harbour Board, Treasurer of the University of Liverpool

It has long been held to be a truism in Political Economy that price regulates production. In many respects this view seems to need serious modification as regards the subject with which we are dealing. This seems to be so not only in this country, but also in those other countries from which we are accustomed to draw our supplies.

It should be clearly understood that any generalisations indicated here are independent of present war conditions. The position of this country as regards wheat supplies is very much in our thoughts, and is giving rise, no doubt, to serious consideration. It is quite possible that some of my readers may be tempted to come to certain conclusions based on our present supplies. Such conclusions may, or may not, prove in the future to have been wise ones. In any case, conditions are now so abnormal that any speculations based on them or even influenced by them are to be deprecated.

While endeavouring to avoid, as far as possible, figures of crops, acreages, yields of crops, etc., it will be necessary to refer occasionally to figures, especially by way of comparison.

Since this paper was prepared certain very important proposals have been brought before Parliament with a view to stimulating British farming.[1] While it is not possible completely to ignore them, any observations must be taken as entirely unofficial. There will be no attempt to discuss the necessity or desirability of the measures proposed, but merely to indicate some of the questions which will inevitably arise in carrying them into effect. The aim has been to deal with conditions as they were before the war, and as they are likely to be after the war is over and its immediate effects have passed away.

No attempt is made to show that the Empire is or might be made self-supporting. Hence the subject chosen is "The Wheat Supplies of the British Isles," rather than of the Empire.

Before coming to the subject proper, a word or two must be said about flour-milling, especially about its history and development during the last part of the nineteenth century. This development, by its influence on distribution of wheat, appears to have had a most serious effect on wheat production in this country.

If we examine the published tables of the growth of wheat in this country during the last seventy years, we find that since 1868, the year of the last big crop of English wheat, when under exceptionally favourable climatic conditions, a crop of 16½ million quarters was raised, a steady and, in fact, rapid decline in the acreage

[1] This paper was written in February 1917, before the Corn Production Bill was introduced into Parliament.

of English wheat has taken place concurrently with a steady increase in population. Thus in the year 1868 the acreage under wheat was all but 4,000,000, while the population was 30,000,000; to-day, with a population of 47,000,000, the acreage has fallen to about 1,600,000. During the last fifty years the population has increased by more than 50 per cent., while the acreage has decreased 60 per cent.

These are rather startling figures. Tempting as it would be to investigate all the probable causes which have so reduced our home supply of wheat while there has been such a steady increase in the demand, it is only possible here to refer to what have been the two main reasons. In the first place, this decrease has been brought about by the rapid fall in the value of wheat owing to cheap ocean freight and the virgin soils of more or less unlimited extent, which were so rapidly brought into cultivation. The second cause may seem somewhat paradoxical: it is apparently the steady and continuous increase in the population of the British Isles, especially in the manufacturing districts.

Returning again to the year 1868, we find that on a large acreage, and, as far as wheat was concerned—an ideal spring and summer—it never rained from April till September—a total crop of 16½ million quarters was raised. This was only occasionally exceeded in previous years on rather larger acreages, and never again approached. Yet even in 1868 we had to import 8,000,000 quarters, or nearly one-third of our total requirements. A few years later we were compelled, owing to increased population and decreased production due to less favourable weather, although the acreage was nearly as large, to import just 50 per cent. of our requirements.

In the meantime the world's price of wheat was seeking a lower level, and by 1880 the fall undoubtedly had had a great effect on reducing the amount of land in Great Britain seeded to wheat; the acreage had, in fact, fallen to 3,000,000, a fall of 25 per cent. since 1868, and by this time we were importing nearly two-thirds of our requirements.

It was obvious even in 1880, when the population was still much below what it is now, that we should, quite apart from price, have to look for the bulk of our supply to imported wheat, and this became more and more evident as each year added a substantial quota to our population. Just at this period a complete change in milling machinery was being introduced all over the world. It became apparent to the milling industry of this country that, as it would be compelled to draw, at least, two-thirds of its supplies in future from overseas, the proper place for the mills would be the seaboard instead of scattered up and down the country in convenient places for home-grown wheat.

This change was all the more rapidly effected because so much machinery had to be scrapped, and in the modern mill it is the machinery, not the building, which costs so much. Therefore the closing down of numerous country mills and the establishment of one big mill after another at the seaboard was prompted both by the change in the art of milling and the large imports of foreign wheat.

This new development began in the late 'eighties and assumed large proportions in the 'nineties, just at a time when wheat fell to its lowest point. It is true that the lowest acreage was touched in the year 1895, after the phenomenally low prices of 1894, but the acreage rose considerably in the last years of the nineteenth century, only to fall again severely since. It does not appear that this renewed reduction in acreage was due to the international price, so much as to the continued migration of the miller from inland places to the seaboard. The changed conditions were most noticeable at first in the northern counties of Lancashire and Yorkshire. It was not until the beginning of this century that it spread to the south, and especially to London. The rise of the port miller in London, as distinguished from the small local miller in the Home Counties, has certainly not stimulated the growth of English wheat. The large volume of imports, which have not been checked by the steady rise in price in recent years, have primarily been due to the great increase in population. Our present population necessitates a total supply of wheat alone of 35,000,000 quarters, and with all our other food requirements, this supply will have apparently to be found largely by imports.

Some further improvement in yield by more scientific farming may be brought about if the price warrants the cost, but with the exception of small countries like Belgium, the British average yield is well above that of other importing countries. It is much above the big exporting countries, like Canada and the United States, and the difference is still more marked when compared with such poor average yields as those in India and Australia.

It must also be remembered that our big population

does not live on bread alone. Dairy farming must always be largely carried on at home, for which pasture and hay crops are indispensable. Our hay crop alone at normal prices represents a value of some £80,000,000. Apart from roots, there are large demands for other cereals, many of which, for climatic and transport reasons, are more suitable for home production; thus oats are more hardy than wheat and, being much more bulky, are very much more expensive in the matter of transport. Oats, too, are largely used on the farms, whereas wheat has to be transported to the mill, and this brings me to the inland freight difficulty. Railway carriage of wheat in this country for long distances is almost prohibitive, and for this reason, as it has been pointed out, the milling industry has more and more gravitated to the seaboard. Under normal ocean freight conditions it is generally cheaper to move grain from Chicago to Liverpool than to carry it from Lincolnshire to Birmingham. Those who have studied the railway question in the matter of remuneration of long hauls will know that rates are often lower for the long haul, including a long ocean voyage, than they are for a small part of the haul in this country. Thus it may be cheaper to bring grain from New York to Birmingham *via* Bristol than it is to take grain from the Bristol neighbourhood to Birmingham. On the face of it this differentiation in favour of the longer haul often seems unfair and illogical, but there is much to be said on the other side. It is a highly complicated question, and can only be alluded to here as one of the difficulties our internal traffic has to face.

The labour problem has also, no doubt, had its effect

in hindering the increase of home production of wheat which the improvement in price would have been expected to bring about. The much higher wages in the towns have, no doubt, depleted the labour supply of the country, and it is probable that the farmers themselves are partly to blame in not recognising the coming change, and meeting it by a general rise in wages.

It must be remembered that we are dealing with pre-war times, for no useful purpose can be served by investigating the present altogether abnormal conditions. The labour shortage on the farms may in the future be mitigated by the employment of more mechanical appliances in the preparation of the land and in the gathering of the crops; but our system of small farms, and especially of small fields, does not lend itself so readily as do the prairies of North America to the use of machinery on a large scale.

However desirable it might be for Britain to be self-supporting in the matter of wheat, it is obvious that with her population and requirements of other food-stuffs such a consummation is entirely out of the question. And we are, therefore, compelled to look for the larger portion of our supplies to outside sources.

It is interesting in this connection to study the geographical question as a whole. We are immediately struck with the minuteness of the British Isles on the map of the world; we are further struck with the sparseness of the population per square mile in all the great exporting countries, with the exception of India, as compared with our own. Thus, to take one instance, England and Wales could be absorbed by the state of Kansas alone and leave a margin all round almost equal to half our superficial area. And this state of Kansas often grows for export to Europe and other states in America a crop of 12,000,000 quarters.

Of the countries from which we have been accustomed to draw large supplies, and from which we shall probably continue to do so, it is rather natural in our inquiry to turn first to that great wheat exporting country, the United States of America. At one time it seemed likely to those who knew the United States that the supplies which that country would be able to spare would from the beginning of this century tend to diminish. It was argued that the rapid growth of the population would naturally reduce the surplus, and that the acreage under wheat on both the west and east coasts would, by this very increase of population, tend to decrease, as other foodstuffs would be urgently required. On the east coast it was argued that more and more land would be required for market gardening, dairy farming and the like; and on the western side it was urged that fruit would drive out wheat farming. Both these anticipations have proved to be correct, the New England states have all ceased to be exporters of wheat, and California, which used to supply this country with a large and most important kind of wheat, has quite ceased to export, and, in fact, gets a fair amount of her requirements from other states. But notwithstanding this, the United States crop still remains the most important regular supply that this country has, due partly to bigger acreage in the middle west, and in such states as Kansas and Oklahoma, but also largely to better farming and better yields. Doubtless the higher range of prices has had much to do with this.

In 1894, when the lowest level in wheat prices was reached, the middle west was in sore straits; all farms were mortgaged up to the hilt, and prices were so low that it often did not pay to harvest the grain. There was great tension between the west and the east, the mortgages were largely held by the rich New York bankers, and people freely talked of a separation of the west from the east. There was great agitation for the unlimited coining of silver at a fixed rate compared with gold, and many traders and others, some in very high positions in this country, as well as in America, advocated the adoption of bimetallism. Their argument was, that as there was not enough gold in the world for the world's business, the purchasing power of the sovereign had increased to an altogether unreasonable extent, or, as stated in more ordinary terms, the value of all commodities had fallen abnormally as compared with gold. Just at this time the gold fields in South Africa were being developed, and there is no doubt that the rapid output of gold from that country had much to do with the fall in the purchasing power of the sovereign, and the rise in the price of raw material. A great fillip was given to this rise in 1897–98, due to the failure of the crops in western Europe, and the United States benefited enormously by the high prices which ruled all through that winter. The farmers made very high prices for all the products, and it is said that in 1898 in Kansas City and other western cities many joy fires were lit in the public places fed by the mortgage deeds returned from the New York bankers.

The wheat-growers in America ever since the commencement of this century have been very prosperous, and have been able to greatly improve their standard of farming. The yield of wheat in the United States is now about twenty per cent. better than it was in the last decade of the century.

The near proximity of the United States to this country, the network of railways under very excellent management capable of handling expeditiously and cheaply immense masses of grain, the storage facilities all along the railways, and during the spring and summer months the excellent lake and canal systems, are sure to enable this country to export to Britain very freely as long as the price of wheat pays the cultivator. In addition the United States grows two crops of wheat: one, the principal one, sown in the autumn and known as the winter crop, and the other the spring crop. It is not often that both these crops fail in the same year, consequently even when a severe disaster overtakes one crop there is almost always a considerable surplus available for export. It would appear likely that for some years to come improved farming is likely to offset the increase of population in the United States, and that there is likely to be in normal years a considerable surplus for export available for Great Britain and other importing countries.

In the neighbouring country of Canada the development was for a good many years both slow and uncertain. Much was said twenty years ago about Canada being the granary of the Empire, but in those days the maximum amount available was comparatively small and not infrequently there was a partial failure. For some time also the trunk lines which were constructed early in the history of Canada's westward development had not the necessary feeders in the shape of branch lines, and were consequently unable to do for Canada what the railways in Argentine were then doing for the granary in South America. Up to the end of last century, and even later, the total export from Canada only amounted to two, or occasionally three, million quarters, a very trifling amount compared with our requirements. It was not, indeed, until 1911 that the Canadian crop really became a serious competitor in the markets of western Europe, when their total exports all but reached ten million quarters. By this time the branch lines of the great trunk lines of Canada were bringing new prairie-land into wheat cultivation, and the great province of Saskatchewan was rapidly rising to the first place in the wheat production of the Canadian provinces.

It had always been known that the soil in these western lands was highly suitable for the production of wheat, the difficulty at first was transport and also labour. The transit problem has been largely met by the great railway systems, and mechanical apparatus is enabling the Canadian farmer to do with much less hand labour —the flatness of vast territories and the huge size of the farms encourage the use of labour-saving apparatus, and the haulage of long and heavily laden trains is comparatively easy. In bringing their supplies to the seaboard Canada, like the United States, has also the great advantage of lake, canal and river traffic. The one great disadvantage from which Canada suffers is her long winter and the necessity of the great bulk of her crop being sown in the spring. Spring crops are more precarious than winter crops, and in Canada in addition to the drawbacks of late springs and dry weather in May and June, after seeding has been effected, there is the danger of an early frost in August catching the crop before it has ripened. If the crop is late in being sown this danger is all the greater, and even if a late crop escapes frost it is sometimes partially lost by early autumn rains and snows.

However, experience shows that usually after a new country has been opened up by railways, towns built and clearances effected, the weather conditions become more settled, and in this respect, even though much of its wheat land lies further north than we used to think suitable for wheat, Canada does not appear likely to be an exception. It must not be supposed that there are no exceptions to this generalisation. Unfortunately, there are, but over a reasonable period of years it will be found that bad failures to crops become rare and weather conditions less uncertain.

From the early years of this century the increase of acreage has been steadily growing, until in 1915 a crop of enormous volume was raised which will probably be known for many years as the freak crop of Canada. It resulted in a total export of 30,000,000 quarters, or more than our total requirements of foreign wheat.

This crop was the result of ideal weather conditions. It was said that there was never five minutes of unfavourable weather from start to finish, and the whole of July and August, when so much harm can be done in a night or two, was absolutely ideal. The yields on some of the best lands were consequently phenomenal, entirely eclipsing even the best results in England.

Although such conditions may not recur again for many years, it is quite evident that Canada will be one of the chief sources of supply of this country for the future. Owing to her small population, she may eclipse her neighbour, the United States, and, in fact, become the largest wheat exporter of the world.

The wheat development in the Argentine is a curious one. From the time that this country began to be a serious exporter of wheat, say in 1890, the increase of acreage was not only rapid but continuous. Not so, however, the crops, which fluctuated very much from year to year, culminating in 1897 in a practical failure as far as export was concerned.

The low prices in the early 'nineties, which certainly affected the production of wheat in many of the other exporting countries as well as in our own country, had no effect in Argentina, and this for a curious reason. The great set-back in the development of Argentina which followed the Baring collapse in 1890, resulted in a most serious depreciation of the Argentine currency. The gold premium rose to 450 per cent., and consequently the Argentine farmer, as long as he did not want to purchase goods that were imported from foreign countries, was better off than he had been in the days of higher prices before the great depreciation of his paper currency. He appears to have cultivated his land in a rather extravagant manner, taking all he could get out of the virgin soil and putting very little back. With the increase of acreage his totals steadily rose, notwithstanding several partial failures, until the climax was reached by an export in the crop year of 1907–8 of over sixteen million quarters. This total has never again been reached, and, in fact, only once has it been approached—in the year 1912–13.

The Argentine now has been subjected for several years to great climatic changes, which have always been unfavourable to wheat, though on several occasions of great benefit to maize. It would, indeed, almost appear as though the country were not destined to remain a great wheat exporter. The rival claims of stock-raising and Indian corn-growing, added to the very serious damage which weather, both in seed time and harvest, can in the Argentine inflict on wheat, may result in it permanently occupying a secondary place in the exporting world of wheat.

Of the three big countries which have become large exporters of wheat in recent times, *i. e.* since the year 1890 — Canada, Argentina, Australia — Australia has been the most fickle and uncertain.

In Australia the rise in price has undoubtedly had an important effect in stimulating the production, but the climatic conditions are such that there have been in the last twenty years no less than five complete failures of the crops, and on four occasions it has been necessary to import wheat into Australia for home consumption.

The first considerable export took place in 1904, and the largest of the pre-war crops was shipped in 1914. In 1894 the amount exported from the crop which was gathered in the previous December was 1,000,000 quarters, which had risen to an export in 1904 of nearly 5,000,000 quarters. This does not compare unfavourably with the development in the same period in Canada. But from this year onwards the growth was much slower than that of Canada. In 1914 the shipments amounted to 8,000,000 quarters, or only about half of the Canadian total. In 1915 there were no exports as the crop raised in 1914 proved once more a complete failure, owing to a prolonged period of drought. But the crops of 1915 and 1916, favoured by good weather conditions, have, on a much increased acreage, been unprecedentedly bountiful. This increase of acreage, as is well known, has been directly brought about by the guarantee given by the Australian Government of a certain price, a remunerative price, to the farmers.

Whether, as more land in Australia is broken up, there will be any real and permanent effect on the climate remains to be seen; if so, the Australian crop will undoubtedly become a very valuable source of supply. It is, perhaps, the finest wheat in the world, and from the point of view of consumption in this country, has only one drawback—the long voyage. Apart from the fine quality of flour of excellent flavour, which is produced from this wheat, it is, as the miller describes it, a very "thirsty" wheat; it absorbs a large proportion of water, which the flour retains in large measure even after it is manufactured into bread.

Another very thirsty wheat is that grown in India, but this wheat is not of such good flavour as Australian, nor is it possible to extract quite such a long length of flour out of it. Still, it is a very favourite wheat, especially in this country, and mixes very well with English. It is, indeed, most valuable for this purpose, when English has been gathered in a wet harvest.

The exports from India, perhaps, more than those of any of the other countries we have been examining, are more difficult to forecast from year to year, or even over a period of years. The climatic conditions are a most important factor, especially in the unirrigated provinces, and the price element is of immense importance. But unlike other wheat-exporting countries, the rise or fall in price does not seem to materially affect the acreage. It has varied very much from year to year, but its variation appears to have been mainly due to weather conditions rather than to price movements. Still, there has been a general increase in the acreage during the last twenty years, but this has not more than kept pace with the rise in the standard of living in India. Thus, the exports from India have not often exceeded the 7,000,000 quarters of 1891–92, and only once to any material extent, when in 1904–5 they amounted to 10,000,000 quarters. In addition to price and climatic conditions affecting exports from India, it must also be remembered that in India the psychological equation counts for more than in any other country in matters commercial. The Indian native is a past master at smelling out a rise or fall in the market. Exports often take place from India, which, according to the most carefully compiled statistics, should never do so. Conversely, exports which appear to be almost a mathematical certainty fail to materialise.

On some occasions the native in holding back supplies is not influenced by a rising market, but by some obscure reasons, which lead him to hoard wheat just as he hoards rupees and jewels.[1]

As has been mentioned, the standard of living in India, like that in many other countries, has risen in

[1] Wheat in India will keep perfectly for years if "pitted," that is, stored in holes in the ground.

recent years very markedly, and all the old landmarks as to prices checking wheat consumption in India have been swept away. At present it does not appear likely that India will rival Canada or the United States, or even Australia, in the matter of export. But the supply is likely to continue quite considerable, and will always be very valuable to this country, as it is mainly shipped in the spring and early summer, and arrives during the three summer months before the crops of the northern hemisphere are available for milling.

We now come to the last great source of supply from which this country can in normal times obtain a portion of her wheat requirements. In volume it is probably potentially the largest of all—it is certainly the oldest of all; but is not so generally popular with the British millers as with those of the other importing countries of Europe.

Long before Great Britain was an importer of wheat, Russia and the Balkan Provinces were exporters—and, on occasion, very large exporters. The Balkans are considered along with Russia (including Siberia) because Black Sea shipments are generally classed together for convenience, and, moreover, certain climatic conditions are more or less common to them all.

A word or two will suffice for the Balkans. At present Balkan wheat is not, in the opinion of the British miller, valuable for his purpose. It lacks "strength," is not suitable for mixing with English, and the farming in the Balkans being primitive the wheat contains a large percentage of foreign matter, some of which it is difficult to extract. It does not produce the quality of flour called for in Britain, and the bulk of it goes to the Low Countries to be used—partly there, but very largely in Germany. The crops of both Rumania and Bulgaria are very uncertain, the farming, as has been said, is primitive, and as the result depends on nature's fertilisers—rain and sun at the proper times—the exports to this country never assume large proportions.

In the case of Russia we are less able to form estimates as to future possibilities of supply, than in that of any of the other great sources. We are without any reliable trade information, and what governmental information is available is not of much value. There is a general belief that a huge increase in production is possible, and the future of Russian wheat is well worth the pains of careful first-hand research.

The system of farming in Russia had, in the years immediately preceding the war, been undergoing radical changes. Up till then, large tracts of land were not only cultivated by individual peasants, but since the freeing of the serfs were controlled by them as members of the village communities. Various causes, such as customs of inheritance and difference in the fertility of the soil, frequently led to the area of land cultivated by each individual being in the form of numerous narrow strips, often at considerable distances. It is obvious that such an arrangement involved great waste of labour, in addition to the fact that the good farmer was at the mercy of idle or incompetent neighbours in the matter of weeds. Moreover, the contiguity of other cereals also militated against the production of clean wheat. Quite recently a change in the system of cultivation was initiated by the Government, under which the peasants' holdings were being consolidated. It was hoped that much better farming would result, with a large increase in all cereal crops, especially of wheat. So far there are no reliable statistics to show how far this change has been general, and no doubt some years must elapse before the expected increase makes itself felt. Nevertheless, the fact remains that there are great tracts of land in Russia admirably fitted for wheat-growing, while the large rivers running north and south, particularly those running south, provide great natural advantages for distribution, advantages absent in such a country as Australia. Railways are still in the initial stage, while it frequently happens that crops cannot be moved owing to the impassable condition of the so-called roads. So inefficient are the means of intercommunication between one province and another, that it not seldom happens that a famine is raging in some one part of Russia, while the country as a whole is exporting on a considerable scale.

Here, as elsewhere, climate plays an important part, particularly as regards the spring crop, which in Russia amounts to something like two-thirds of the whole. It is liable to great injury from scorching winds in June, but that month once safely over the spring crop is usually good. The winter snows generally provide the soil with moisture sufficient to ensure satisfactory growth in the early stages. Winter "killing" is not unknown, but here, again, the snow is a benefactor by protecting the crop, which in any case is of less importance compared with the much larger spring crop.

Russian wheat is thoroughly suitable for milling

in this country. The spring wheat possesses great "strength," some of the better sorts equalling Manitoba grain in this respect; while the finest winter varieties, especially those grown in Podolia, combine considerable "strength" with the nutty flavour of the Australian product. The limited supply that has hitherto reached this country is due to the nearness of Italy and the other importing countries of the Mediterranean, and to the fact that Germany and Austria are actually contiguous.

In this somewhat hurried review of the sources of supply available for this country, all the big exporters have been included, whether inside the Empire or out. The problem of wheat supply really concerns the British Isles, rather than the Empire as a whole, which, apart from these islands is either self-supporting, or actually has wheat to spare. The requirements of South Africa, the West Indies, and the Crown Colonies, are relatively so small as not to affect the problem.

Having reviewed the main overseas sources of supply in normal times, we must very briefly consider the probable effects of the Government proposals relating to British farming on the actual wheat-production at home.

There can be no doubt that such a guarantee of prices as it is proposed to give to the farmers will increase the acreage under wheat in the British Isles, and will also encourage more intensive farming, and a corresponding increase in yield from less productive land. Fresh capital will tend to be introduced, and mechanical means of cultivation and harvesting will be adopted. Even without the compulsory enactment of much higher wages for farm hands, the results of the other proposals would probably have brought higher wages in their train. Further, the prohibition of an increase in rent will, no doubt, result, as soon as labour again moves freely, in the gravitation of a considerable proportion of the population to the land, amongst them probably a new type of farmer, more ready to apply new methods. Under these conditions it is quite possible that, instead of growing only one-fifth of our requirements, we may soon be growing a fourth, or possibly even a third. Assuming that this is the case, we come face to face with the distribution problem. Much of the land likely to be newly laid down to wheat will be a long way from the milling centres now so largely on the seaboard. It is obvious that for such bulky goods as grain, water-borne traffic is much cheaper than rail; but, unfortunately, owing to our totally inadequate canal system, water-borne traffic, for us, is chiefly seaborne.

A Commission was appointed some years ago to inquire into this problem. Their investigations proved its urgency, and also that its solution could only be attained if undertaken by the nation as a whole. The policy of national allowing, and in some cases compelling, the railways to control or buy up the canals, has left the present generation a problem bristling with difficulties—water-rights; questions of rates and competition with railways; difficulties of approach and bridges; the impossibility of widening or deepening canals, where they pass through populous districts, without diverting their courses; vested interests of all kinds;—such are some aspects of the problem. In addition, there is the ever-increasing difficulty of providing sufficient storage of water to meet the wastage consequent on the number of locks involved on the heavy gradients obtaining in many parts of the country.

We move very quickly in these war-times, and it is interesting to note that the main lines of canal communication have been taken over by the State. They are being worked in conjunction with the railways, and so Mr. Neville Chamberlain recently expressed his belief at Sheffield, that that unification of control will never again be destroyed.

This pronouncement is of the happiest augury for the future of our commercial distribution, and it will be of the utmost benefit if a great scheme of canal enlargement and consolidation under State control, can be effected. It will go far to solve the difficulty of bringing wheat at a moderate price from the wheat districts to the milling centres.

It is true that the transport difficulty might further be met, by bringing the mills to the grain, instead of the grain to the mills. But it would be an expensive operation, and there would still be the problem of carrying the flour to the thickly populated districts, often a long way from wheat-growing land. Moreover, the flour milled on the spot would be mainly, if not entirely, of English wheat, as it would not pay the miller to bring large quantities of foreign wheat to the country districts, and have to move it once more, in the form of flour, to the great centres of population.

In years when weather at harvest-time is unpropitious, flour made from English wheat alone would compete disadvantageously with port-milled flour, in which foreign wheat is used in considerable proportions, and this brings us to the weather problem. It is true that our wheat crop is rarely affected adversely by weather during the period of growth—winter-kill is almost unknown, and the effects of a prolonged spring drought are generally counteracted by the power of the wheat to send down a long tap-root in search of moisture. No means whatever have yet been devised of overcoming the ill-results of bad weather, except the makeshift of stacking the wheat and then waiting for cold weather to put it into condition. This delay would be a serious difficulty for the inland miller, if compelled to mill English wheat alone.

The whole subject of English wheat production is a fascinating one, which it behoves us all to consider in the light of new conditions and experience, and which now, more than ever, will repay the investigator. He has to consider how far acreage is affected by questions of price and of distribution (a most complicated matter as we have seen), and how far it is virtually settled by the claims of certain rival crops which are essential for our huge population, and which must be home-grown.

Though the financial issue raised by the proposed arrangements with farmers in this country cannot be dealt with here, its bearing on our overseas supplies should not be forgotten, supplies which can obviously be grown at a much lower price than that proposed for British produce. It is, however, clear, that with a largely increased acreage at home, the Empire might be self-supporting (though at considerably increased cost) even in the eventuality of one or other of our Imperial sources of supply proving a complete failure. That such a failure is a contingency for which we should be prepared is evident from our past experience. Though there can be no sort of doubt that, in times like the present world crisis, it would be a great advantage for the Empire to be self-supporting, opinions differ widely as to the wisdom of a policy which aims at a self-contained empire in normal times. In any case, we shall all agree that the decision is one for politicians rather than for economists, and that it cannot therefore be considered in a lecture such as this.

THE IMPORTANCE OF IMPERIAL WOOL

By E. P. HITCHCOCK, M.A.,
Raw Materials Department of the War Office

THERE are few commodities so intricate and diverse in their nature as wool. Other textiles, such as cotton and flax, can readily be standardised, but in the case of raw wool this is extremely difficult. Questions of yield —the amount of clean wool which can be obtained from raw wool from the sheep's back after abstracting grease, moisture and other foreign matter which it contains— the quality, the style and the length of the fibre are all factors rendering wool one of the most difficult raw materials to buy and to sell. Raw wool from one locality or from a particular breed of sheep may produce, when it is worked up, only one-fifth of its original weight, whilst in another district or from another breed it may yield four times this amount.

So intricate is this matter of yield that even the farmer who grows the wool is very rarely able to value it; even from the same sheep wool will vary in its yield, and therefore in its value, from year to year, owing solely to changes in the weather. This diversity in the matter of yield renders wool statistics most misleading, for they are generally given in terms of greasy wool. All wool figures should be treated with considerable reserve and subject always to modifications on technical grounds.

Wool may be roughly divided into two classes: the

first consisting of the finer kinds, from which clothing is produced; and the second of the very coarse varieties used for carpets, felts and for the coarsest clothing materials. Roughly, two-thirds of the world's wool is suitable for the manufacture of clothing, and one-third for carpets, etc.

The finer wools may be subdivided into two classes: the very fine fibre "merino" and the coarser fibre "crossbred." It is the practice to classify wool in qualities (30's to 58's for crossbred, and 60's and upwards merino), which originally corresponded to the "counts" to which they would spin, but in the course of time this has ceased to be the case. But no very exact and definite classification is possible, since fine crossbreds merge almost imperceptibly into the coarser merinos.

An examination into the distribution of merino and crossbred sheep throughout the world, for the last twenty years, shows that there has been a gradual but sure tendency for crossbred to displace merino sheep. The reason is not far to seek. The crossbred sheep provides mutton, and with the increase in freezing facilities the rearing of sheep for their mutton as well as their wool has become a profitable business. Australia, where at least 60 per cent. of the British Empire wool is grown, at one time produced only merino sheep, but now has 20 per cent. of the crossbred variety, a proportion which is increasing Of the New Zealand flock 97 per cent. are crossbred; in the United Kingdom there are no merinos, while in South Africa there are no crossbred varieties.

Before the war, the world's total production of merino

and crossbred wool was roughly 2000 million lbs. divided almost equally into merino and crossbred, and the carpet wool production about 650 million lbs. One-third of the world's merino and crossbred wool is produced in industrial countries, which consume it themselves; the remaining two-thirds is exported from pastoral countries. It is mainly with this exportable wool that the following pages deal.

Practically every industrial country produces wool. Europe, which works up the bulk of the world's supply, is, in fact, the densest sheep and wool producing continent in the world, but every industrial country must import in addition wool for purposes of manufacture, not only to augment quantity, but also quality, for in order to produce various fabrics, wools of different descriptions have to be mixed or "blended." All industrial countries, therefore, whether they produce wool or not, find it necessary to import large quantities.

London is the world's greatest wool market, though of late years there has been a tendency for consumers to purchase direct in primary markets, as the following figures will show—

	Percentage of Colonial Clip bought at London Sales.	Percentage of Colonial Clip bought direct in Colonies.
1895	63	37
1894–8	59	41
1913–14	27	73
1915–16	29	71

Simplification of handling arrangements and a more economical system of labour organisation at the London

Docks may, however, have some effect after the war in stemming the already noticeable decline in the importance of London as the world's wool market.

The chief crossbred and merino wool exporting countries can be roughly divided into two geographical groups: the British Empire group consisting of Australia, New Zealand, and South Africa, and the South American group including Argentine, Uruguay and the southern part of South America, known popularly as Patagonia, and among wool men as "Puntas," from the port of Puntas Arenas in the southernmost point of Chili, the island of Tierra del Fuego, through which Patagonian wool is mainly shipped. The Falkland Islands also export a small amount of wool. The production of the British Empire group (excluding the United Kingdom) is about double that of South-America, and it includes 80 per cent. of the world's exportable merino wool, of which the Australian variety is the finest produced. It is, therefore, clear that in the production of wool the British Empire holds a commanding position, and has within it a virtual monopoly of the supplies of the world. It has, moreover, considerable possibilities of further expansion of production. Of the British Empire production three-eighths is crossbred and five-eighths merino, and of the South American clip almost one-third is merino and the rest crossbred.

The amount of wool produced in a country depends upon three main factors, the number of sheep, the weight of the fleece and the yield of the wool. Sheep population is generally a somewhat inadequate indication of wool production—the weight and yield of the fleece are the important factors. In some parts of South Africa wool yields as low as 18 per cent., while on the other hand some varieties scoured in South Africa yield in this country up to 105 per cent.—moisture having been absorbed by the wool during the voyage. English wool yields 70 to 85 per cent., whilst crossbred generally yields from 60 to 70 per cent., and merino 35 to 45 per cent.

The following figures give some approximate idea of the distribution of wool production throughout the world.

Europe had before the war, fifty sheep to the square mile, producing about 212 lbs. Australasia has now thirty-one sheep to the square mile, producing 211 lbs. of wool.[1] South American ranks next, with sixteen sheep to the square mile, and 100 lbs. of wool. In North America there are not more than twelve sheep to the square mile, and 58 lbs. of wool, while in Africa there are six sheep to the square mile and 20 lbs. of wool. Asia, which was the original home of the sheep, has now only six to the square mile and 16 lbs. of wool.

Before the war Europe produced more wool than Australia and New Zealand put together. The war has been responsible for a considerable decrease, mainly owing to slaughtering of sheep for food, and we shall have to notice later the effect of this upon the probable world wool position after the war. At one time the wool production of Germany was very important, in

[1] The number of sheep per square mile in New Zealand is 233, and 1876 lbs. of wool are produced. The number of sheep per square mile in the United Kingdom is 228, whilst the weight of wool produced is 1008 lbs., the bulk of it in a washed state, whilst in New Zealand the wool is greasy.

quality as in quantity, but it has for many years been rapidly decreasing; while the United Kingdom has about twenty-eight million sheep, Germany before the war had but five million. It is interesting to remember that in the past Germany relied upon the British Empire for over 50 per cent. of her supplies, mainly of merino wools. She has of recent years made strenuous efforts to increase the wool production of her colonies, particularly those in East Africa, where at the beginning of the war there were more sheep than in the whole of Germany, but the supply of raw wool in the future for her textile industry is a matter of serious concern to her. At a recent meeting of German agrarian interests convened by the State, it was seriously proposed that, in order to stimulate wool production in Germany, minimum prices for wool for the next fifteen to twenty years should be State guaranteed. But not only Germany is taking steps to increase her wool production. Apart from the automatic incentive to produce more wool which has resulted from the high prices of the last three years, production is being fostered by manufacturing interests in various countries, notably in America.

Wool differs in many characteristics from cotton. If the cotton crop of the world is low one year, high prices will probably lead to an increased acreage being brought under cultivation, with the result that the crop considerably increases the following season. That is not the case with wool. The generation of a sheep is about four years, and if there is any reduction in production it cannot be remedied under a number of years.

It happens that the great wool manufacturing areas

of France, Belgium and the Lodz district of Russia are all near the frontiers. With the advance of the invading German armies on all frontiers, the wool industry of the Continent was, therefore, affected more than any of the other great continental industries. August, 1914, was a month of dismal prophecies. With the practical cessation of the Continental demand for wool, stagnation was predicted for the wool trade, since in pre-war years the Continent had taken no less than 60 per cent. of the British Empire and 79 per cent. of the South American wool production. It was expected that some relief would be given by the demands for clothing the Allied armies, but that this would only affect crossbred wool, for merino was little used for the production of army cloth. Business was almost at standstill, and the trade generally regarded a very serious fall in prices as inevitable, and many predicted ruin for all but the strongest wool men.

The forecasts have, however, been contradicted by the subsequent facts. All previous records in prices have been surpassed and again surpassed, and the strongest fact is that, although all wool prices have increased enormously, the rise has been most marked and striking in the case of wool not required for military purposes. A temporary decline in prices took place, but it was due almost entirely to a misapprehension by the trade not only of the effect of the direct war demand, but more especially of the important factor of American competition, which had, in fact, set in at the beginning of the year consequent upon the passing of the Simmons-Underwood Tariff Bill in October, 1913. It is very easy to be long-sighted after the event, but it is clear that even without the stimulus caused by the necessity of providing cloth for the increasing Allied armies the wool market was rising from other more permanent economic forces.

Before considering these factors it is interesting to consider the actual war demand. In normal pre-war days the requirements of the British Army and Navy did not amount to one per cent. of our total wool consumption. Within three months from the beginning of the war it had jumped to over 10 per cent., and since then it has almost equalled it. A civilian in this country consumes normally about 13 lbs. of wool a year, and England is a very heavy wool-consuming country. A soldier's outfit accounts for roughly 32 lbs. of wool, and a soldier does not merely require one outfit per year—but many. The demand made upon raw material and upon the output of the trade have since been unprecedented. In normal years the total consumption of wool in the United Kingdom is about 545 million lbs. and the total yardage produced about 420 million yards. In 1915 the total of the wool imported for consumption was 894 million lbs., and total output 611 million yards. Since the beginning of the war the War Office has placed contracts for woollen goods to a value of over £100,000,000, and has purchased raw wool to the value of about £50,000,000. It has placed orders for over 120,000 miles of cloth. The production of flannel for military purposes during the first eighteen months of war was equal to over two and a half times the annual production for all purposes before the war, and that of blankets five times the pre-war output.

The War Office has also to provide to a large extent

the requirements of the Allies partly in the form of raw wool and partly as cloth. At the beginning of the war the textiles areas of France, of Belgium, of Poland and of the Lodz district of Russia, near the frontier of the Central Empires, were invaded and occupied by the enemy, and incidentally sufficient wool was requisitioned to clothe the German armies for a year. France, who before the war had a magnificent wool industry—both her imports and exports of wool were greater in value than those of any other commodities—lost nine-tenths of this, and is now reduced to one-eighth of her pre-war trade. The Belgian army has had to be re-equipped entirely from England. Before the war the trade of Belgium was very considerable, although 75 per cent. of it consisted in scouring and carbonising on commission for other countries. The Russian Government has organised the mills in the Moscow and other districts to meet the huge requirements of its army, and for the last two and a half years there has been no appreciable production for civil requirements. In spite of this, however, it has been impossible for Russia from her own machinery to produce sufficient woollen cloth for her armies and in spite of the use of a considerable amount of cotton clothing. This country has, therefore, had to produce very large quantities of Russian khaki.

The bulk of the war demand for wool has been focused in the United Kingdom. To meet this it has been necessary for the Government tc organise in addition to the production of cloth, the distribution of the raw material itself and the shipping facilities for bringing it overseas. Following the example of France, who requisitioned the whole of her domestic wool clip in the earlier days of the war, the United Kingdom purchased in 1916 her domestic clip of 100 million lbs., paying the farmers 35 per cent. over 1914 pre-war prices, or 10 per cent. higher than French growers received in the same year. By agreements with the Australian and New Zealand Governments the Australasian clips were purchased in 1916 at prices 55 per cent. above the 1913–14 prices to the growers.

The purely spectacular aspect of the military demand for wool should not, however, blind us to the other more permanent factors which have contributed to the present wool position, and have sent up the prices of merino wool far more than the military demand raised the price of crossbred wool. The two main elements of importance are—

(*a*) A decrease in the world production of wool.
(*b*) A much increased civilian consumption in all countries outside the Continent of Europe, especially the U.S.A. and to a lesser extent the United Kingdom.

The decrease in wool production represents approximately 20 per cent. of the world's exportable wool supplies, equal to over 300 million lbs., and equivalent to the total pre-war import for consumption of the United States of America, Italy, Russia, Japan, Norway, Sweden, Denmark and Holland. The main cause of the decrease was the exceptionally heavy drought experienced in Australia in the 1914–15 season, which reduced the sheep flocks from 85 millions to 68 millions—a decrease of 17 million sheep.

A further contributory cause is to be found in South America, where wool production has been declining for many years owing to the competition of wheat-growing and cattle-raising, which appear to be more profitable than the production of sheep. The wholesale slaughtering of sheep throughout Europe during the war does not appreciably affect the present position, though in any review it must, of course, be taken into consideration. When, after the war, we are once more in a position to take stock of the sheep population of Europe, we shall probably find the wool production of the Continent to have decreased enormously, and this will contribute still further to a world shortage and add to the demand for overseas supplies. When Germany invaded France she took off a great number of sheep, the bulk of which are slaughtered; an abnormal proportion have shared the same fate in Belgium, Germany, Austria, Turkey and in the Balkans generally.

The second factor is the enormously increased demand for civil purposes outside Europe, especially from America, and in a lesser degree from Eastern markets —Japan and China. It has been stated that the increased demand for wool from America has been due to the placing of Allied orders for clothing in that country, but that is incorrect. No such orders have been placed in America for the last two years. The factors making for the increased import of wool into America were in operation before the war started, the greater element being the tariff.

For a considerable period before the war there had been a very heavy import duty—as high as 75 per cent. *ad valorem*—on wool imported into America. In

October, 1913, however, the Simmons-Underwood Tariff Bill introduced a period of free wool imports. In the six months prior to the war American purchases of colonial wool had increased three-fold in comparison with the similar six months of the previous year. Last year the United States imported roughly five times their pre-war quantity, having purchased one-third of the Australian clip and over one-third of the South American clips. Before the war America purchased not more than 1000 to 2000 bales of South African wool, while in the 1915–16 season she took almost 200,000 bales. The increase in popular purchasing power in America following upon the tariff have completely revolutionised the whole world wool position.

The increase in American wool purchases in the British Empire and South America may be seen by the following figures—

	1914.	1915.	1916.
	Per cent.	Per cent.	Per cent.
United Kingdom from British Empire	34	72	58
Continent " "	60	8	11
United States " "	6	20	31
United Kingdom from South Africa	12	20·5	10
Continent " "	77	48·5	43
United States " "	11	31	47

The cessation of Continental demand, equivalent to about 50 per cent. of the colonial clips, has been roughly compensated by the war demands of the Allied armies. The increased demand from America and other countries and the decrease in world production have more than absorbed the balance. As a result stocks of wool have been reduced below the normal, and consumption to-day all over the world is higher than it has ever been. Yet the shelves of the world are almost empty. Manufacturers and merchants in this and other countries who once had very considerable difficulty in finding customers, cannot in many cases guarantee delivery now or even a year hence.

The underlying features of the wool position are not temporary conditions produced by the war; they are economic and permanent. They have been aggravated by the war, and they are likely to persist after the war. It is possible that the destruction of textile machinery in Europe may temporarily reduce the consumption of wool after hostilities cease, and so ease the situation. Textile machinery in Japan, America and Italy has been considerably increased during the war, and the demand of these countries for wool will be greater in future. Even then, however, the mrchinery that does exist will require to be clothed with a stock of wool necessary quite apart from the wool to be consumed. This will take any quantity up to 100 million lbs.

It is interesting to note that after both the American and the Franco-Prussian wars wool prices remained high for some years, and there seems every probability that history will repeat itself again in this respect.

In considering, however, the question of demand for wool, the effect of possible substitution, if prices soar above a certain point, must be borne in mind. No suitable substitute for wool has yet, however, been found. If wool prices are too high, people will turn to inferior clothing. If there is enough cotton, though it is doubtful if the cotton crop of the world will be equal to the demand, people will use that as far as possible, instead of wool. During the last two years Germany has from necessity been trying to find some substitute, and fibres of all descriptions are being utilised, from bark fibre to nettles and seaweed. Paper has been extensively employed, mainly for underclothing, and good results have been obtained. Bullrushes, fibres from bark and from wood, and a seaweed—*Posedonia Australis*—which is found round Heligoland, have been used with wool, shoddy and cotton and other fibres to produce yarn. Though some of these substitutes will in all probability be used after the war, there does not appear to be any prospect of wool being replaced by anything else which is really satisfactory.

We have seen that the British Empire has a potential control of the wool supplies of the world. We have also seen that raw wool supplies after the war are likely to be short and stocks in manufacturers' and merchants' hands very low. The whole of the Russian market has been swept clear of textiles; there are no textiles in Germany or in Belgium, and very little in France. Every consideration appears to indicate that the economic problem after the war will resolve itself not into a question of the capture of markets for manufactured goods, but rather in the acquiring of sufficient raw material to maintain industry. The country that has its raw material assured will have an enormous advantage.

Germany realises the position, and some of her best brains are carefully considering the problem and formulating post-war policy. For war purposes she has had to organise her cartels, to marshal her resources, to requisition every ounce of raw material, to earmark every piece of machinery. She has also taken steps to re-adapt all that machinery to meet the situation after the war. State-aided monopoly import companies, in which trade interests have part control, have been formed, to buy raw material of all descriptions throughout the world to meet Germany's after-war demands. From the early days of the war the policy of forward purchasing of wool has been pursued by Germany in South America. Altogether she has bought roughly 100 million lbs., which has since been reduced to 60 million lbs. This policy has, however, now been stopped owing to difficulties of exchange.

Apart altogether from the success which her present plans for after-war trade may achieve with regard to the requisition of raw material, one point stands out clearly —Germany's future entry into the raw material markets will be in the form of organised state-aided trusts. A recent report of the meeting of a German Industrial Union contained one very striking assertion in this connection. It boasted that after the war, owing to the steps they have taken now, the organisation which they have set up "will enable 120,000,000 people to come on to the world's market as a single purchaser."

In order to safeguard British and Allied manufactures this organisation will have to be met by one form of organisation just as strong and as comprehensive. After the war pre-war methods will have to be re-adapted to meet the new conditions.

The problem before the British Empire is to secure for British and Allied textile machinery after the war a prior claim upon Imperial wool supplies. It is not my business to suggest how this aim might be accomplished. I have stated the problem and will conclude by repeating the relevant resolutions passed at the Paris Conference for the consideration of the Allied Governments.

"B1. The Allies declare their common determination to secure the re-establishment of the countries suffering from acts of destruction, spoliation, and unjust requisition, and decide to join in devising means to secure restoration to those countries, as a prior claim, of their raw materials, industrial and agricultural plant, stock and mercantile fleet; or to assist them to re-equip themselves in these respects."

"B3. The Allies declare themselves agreed to conserve for the Allied countries before all others their natural resources during the whole period of commercial, industrial, agricultural and maritime reconstruction, and for this purpose they undertake to establish special arrangements to facilitate the interchange of these resources."

PRINTED IN GREAT BRITAIN BY RICHARD CLAY & SONS, LIMITED,
BRUNSWICK ST., STAMFORD ST., S.E. 1, AND BUNGAY, SUFFOLK.

Printed in Dunstable, United Kingdom

91220037R00111